4295
68I

ADVANCED
BIOCHEMICAL
ENGINEERING

ADVANCED BIOCHEMICAL ENGINEERING

Edited by

HENRY R. BUNGAY

GEORGES BELFORT

A Wiley-Interscience Publication

JOHN WILEY & SONS

New York / Chichester / Brisbane / Toronto / Singapore

Library of Congress Cataloging in Publication Data:

Advanced biochemical engineering.

"A Wiley-Interscience publication."
Includes index.
1. Biochemical engineering. I. Bungay, Henry R.
(Henry Robert), 1928–   II. Belfort, Georges.
TP248.3.A37 1987   660'.63   86-15834
ISBN 0-471-81279-X

Printed in the United States of America

10 9 8 7 6 5 4 3 2 1

CONTRIBUTORS

GEORGES BELFORT, Professor of Chemical Engineering, Rensselaer Polytechnic Institute, Troy, New York

HENRY R. BUNGAY, Professor of Chemical and Environmental Engineering, Rensselaer Polytechnic Institute, Troy, New York

LENORE S. CLESCERI, Associate Professor of Biology, Rensselaer Polytechnic Institute, Troy, New York

RICHARD P. ELANDER, Vice-president, Fermentation Research and Development, Bristol-Myers Company, Industrial Division, Syracuse, New York

MICHAEL H. HANNA, Associate Professor of Biology, Rensselaer Polytechnic Institute, Troy, New York

DAVID S. HOLMES, Staff Molecular Biologist, General Electric Company, Research and Development Center, Schenectady, New York

CLEMENT KLEINSTREUER, Associate Professor of Mechanical Engineering, Member, Biotechnology Group, North Carolina State University, Raleigh, North Carolina

MICHAEL R. LADISCH, Professor of Agricultural and Chemical Engineering, Group Leader, Laboratory of Renewable Resource Engineering, Purdue University, West Lafayette, Indiana

E. BRUCE NAUMAN, Professor of Chemical Engineering, Rensselaer Polytechnic Institute, Troy, New York

GEORGE T. TSAO, Professor of Chemical Engineering and Director of Laboratory of Renewable Resource Engineering, Purdue University, West Lafayette, Indiana

PREFACE

New intellectual challenges for biochemical engineers have emerged with the recent impressive progress of genetic splicing and cell fusion technology and the parallel advances in the instrumentation and analytical methodologies. These developments have convinced many universities that biotechnology is one of the key areas for research and education in chemical engineering. The Rensselaer program in biochemical engineering emphasizes education in modern biology and in engineering fundamentals. Our research foci are fermentation, mammalian cell culture, bioprocess dynamics, design of bioreactors, membrane separations, and downstream processing. We have invited distinguished authorities to join us each summer to teach a short course entitled Biochemical Engineering: Separations, Fermentation, and Genetics. This book is based on that course and features selected advanced topics in biochemical engineering.

The term "biotechnology" encompasses scientific disciplines such as biochemistry, molecular biology, microbiology, genetics, and immunology, as well as engineering. Although applied genetics, microbiology, and molecular enzyme engineering are included, this book stresses the application of engineering principles. Chapters on bioreactors, biomass refining, and concentration and recovery of bioproducts are not usually covered in this detail in other biochemical engineering texts. Most topics are presented without much introduction, and we assume that the reader has been exposed to fundamental biology, biochemistry, and chemical engineering. The results are some fresh insights in a unique book that complements other textbooks in biochemical engineering.

Since a large amount of the material covered in our academic and summer courses involves computers, bioprocess simulation, and programs for teaching with computers, this material is published in *Computer Games and Simulation for Biochemical Engineering,* Wiley, 1985.

We thank the contributing authors for their encouragement, enthusiasm, and efforts in making this project possible. We also acknowledge our students and colleagues for their assistance in the short course and in preparing this text.

HENRY R. BUNGAY
GEORGES BELFORT

Troy, New York
December 1986

CONTENTS

1 AN OVERVIEW OF BIOCHEMICAL ENGINEERING   1

   E. Bruce Nauman

2 MICROBIAL CELLS AND ENZYMES   13

   Henry R. Bungay, Lenore S. Clesceri,
   and George T. Tsao

3 ANALYSIS OF BIOLOGICAL REACTORS   33

   Clement Kleinstreuer

4 BIOMASS REFINING   79

   George T. Tsao, Michael R. Ladisch,
   and Henry R. Bungay

5 APPLIED GENETICS FOR BIOCHEMICAL ENGINEERING:
   RECOMBINANT DNA   103

   Michael H. Hanna

6   MOLECULAR ENZYME ENGINEERING   129

       David S. Holmes

7   APPLIED GENETICS AND MOLECULAR BIOLOGY OF
     INDUSTRIAL MICROORGANISMS   167

       Richard P. Elander

8   CHALLENGES AND OPPORTUNITIES IN PRODUCT
     RECOVERY   187

       Georges Belfort

9   SEPARATION BY SORPTION   219

       Michael R. Ladisch

10 MEMBRANE SEPARATION TECHNOLOGY:
     AN OVERVIEW   239

       Georges Belfort

INDEX   299

ADVANCED
BIOCHEMICAL
ENGINEERING

1

AN OVERVIEW
OF BIOCHEMICAL
ENGINEERING

E. BRUCE NAUMAN

Professor of Chemical Engineering
Rensselaer Polytechnic Institute
Troy, New York

1. BIOCHEMICAL ENGINEERS

The 1920s and the 1930s saw the birth of the petrochemical industry brought about by the creative union of chemical engineers with organic chemists. The 1980's and 90's will see a biochemical industry, born from a new creative union of biochemical engineers and molecular biologists.

The words have been used to open the R.P.I. summer courses in Biochemical Engineering. They are meant to inspire and also to suggest shared effort between engineers and scientists in the development of biotechnology. The engineering half of the partnership will be biochemical engineers who are chemical engineers with special training in biochemistry, microbiology, and those aspects of conventional chemical engineering having particular relevance to biotechnology. These biochemical engineers need a background in heat and mass balances, thermodynamics, transport phenomena, reaction kinetics, mathematical modeling, and plant design, all of which are part of traditional chemical engineering. They also need greater exposure and experience in the biological sciences than petrochemical engineers need in con-

TABLE 1.1. Profit Potential of Biochemicals

Selling Price	Class of Compounds	Examples	Subsidy-Free Profitability
Low	Fuels	Methane Ethanol	Negative to marginal
	Foods and food supplements	Single-cell proteins	Marginal to good
		Amino acids	
	Chemical intermediates	Itaconic acid Butanediol	Good
	Specialty chemicals	Agricultural chemicals	Good to excellent
High	Pharmaceuticals	Antibiotics Interferon	Excellent if effective

ventional organic chemistry. This greater need in the supporting sciences is caused by the newness of the field and the new partnership being formed with biologists. By way of analogy, the chemical engineers of the 1920s tended to have more exposure to chemistry than they now have.

The scope of biochemical engineering is beginning to include complex structures of biological origin, for example, monoclonal plants for agriculture, just as chemical engineers now participate in the manufacture of fabricated plastic structures and microelectronic components, topics far afield from the old chemical process industries. However, biotechnology is now largely confined to the production of organic chemicals by microbial (or enzymatic) processes. Table 1.1 categorizes these chemical products in terms of their end uses: fuels, foods, chemical intermediates, and pharmaceuticals.

2. BIOCHEMICALS—FUELS

Fuels are at the low selling price end of the biotechnology industry. Methane and ethanol are readily obtained by fermentation of biomass using processes developed during the 1970s. In retrospect, much of the enthusiasm for biomass fuels was perhaps two decades premature. Biologically derived chemicals are now hard pressed to compete against fossil fuels when energy content is the sole determinant of value. Methane generation from biomass can sometimes be justified as a by-product of refuse disposal. There are now more than 40 landfill-gas wells across the country, and this number has been projected to grow to 100 by the end of the decade. However, direct combustion of the refuse usually gives more energy and is currently favored for environmental reasons.

Ethanol from biomass is more likely to be profitable, in part because

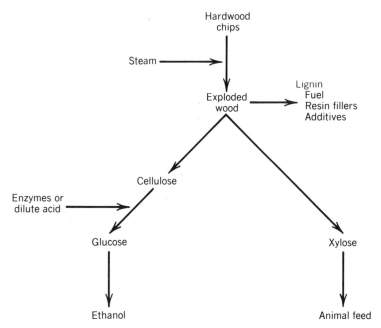

Figure 1.1 Wood sugars from hardwood chips.

ethanol has value as a solvent and antiknock additive for unleaded gasoline. In the United States, ethanol fermented from corn tends to be closely competitive with ethanol from ethylene. Large fermentation plants in the Midwest have benefited from tax subsidies for ethanol as a gasoline additive. Without the subsidy, aromatics or methanol are cheaper. Ethanol from wood-derived sugars is potentially cheaper than ethanol from either corn starch or ethylene. Figure 1.1 shows a typical sugar manufacturing process that gives both five-carbon (xylose) and six-carbon (glucose) sugars. Extensive pilot scale development has been done on processes of this type, and production units are being planned but have not yet been built. The economics require very large plants, over 100 dry tons per day of wood chips, with comensurately large technical and economic risks. Economic improvements will result from the genetically engineered yeast strains capable of fermenting five-carbon sugars. Even with this, however, wood sugar processes will require government subsidies, substantial by-product credits from lignin, or a substantial change in the cost of energy before they can be profitable.

Fuels from biomass are critically dependent on the substrate cost and by-product economics. Much of the value of corn is represented by protein fractions not used in ethanol fermentation, and stillage brings revenue as a cattle feed. The processing cost in going from biomass to finished fuel is likely to represent a major fraction of the fuel's selling price. Good process engineering is thus important, both for the fermentation and for the subsequent separation steps. Ethanol must be concentrated from about 10% in water to nearly 100%. This recovery is conventional chemical enginering in the sense that no biology is involved. However, the problems encountered tend to be rather different from those found in petrochemical plants. For ethanol, the traditional route is distillation (with vapor phase recompression) up to concentrations of 95%, followed by an azeotropic distillation. An attractive option for drying partly purified ethanol by thermal-swing adsorption is discussed in Chapter 4.

3. BIOCHEMICALS—FOODSTUFFS

Too broad a view of foods and food supplements would encompass all of food engineering. Properly included in biochemical engineering are those fermentations and enzymatic conversions that yield foodstuffs. Examples are the production of single-cell protein and individual amino acids by fermentation and the manufacture of fructose using an immobilized enzyme. The amino acid and fructose processes are commercial and presumably have good economics. The market value of these compounds clearly exceeds their inherent energy value. However, their selling prices tend to be limited by competition with naturally occurring foodstuffs. Single-cell protein was conceived as a means of upgrading cheap petrochemicals, methanol, and natural gas liquids into animal feed. Several large-scale plants were built in Italy and Britain, but they have failed or have had marginal profitability. Shifting of prices and regulatory problems have hindered commercialization. The microorganisms used in the fermentation of natural gas liquids concentrate trace aromatics in the feedstocks, and these cause concern even for animals feeds. The methanol fermentation process developed by the English company ICI may be adapted by the Soviet Union as a means of using their cheap natural gas. However, it now seems unlikely that methanol-based single-cell protein will ever be attractive in the free-world economy. The alternative of obtaining single-cell protein (i.e., yeast cells) by fermenting waste streams of biological origin is attractive, however. Whey is now used as a substrate for yeast production. The yeast is used for animal feed.

Biochemicals used directly for human consumption include individual amino acids, fructose, and citric acid for which natural, nonfermentation substitutes exist. Some amino acids can also be made petrochemically. In most cases, the product value is determined by the natural sources rather than alternative synthesis routes. A few specialities such as monosodium glutamate and aspartame have no direct substitutes. Presumably, the economics of these specialities are quite good. In general, however, biochemicals used as foodstuffs tend to be less valuable than the more strict chemical products listed as organic intermediates in Table 1.1. They also face greater requirements for regulatory testing.

4. BIOCHEMICALS—ORGANIC INTERMEDIATES

Relatively few organic intermediates are now made by fermentations or enzymatic conversions, but this class of products holds great future promise. Butanol and coproduct acetone were made by fermentation for more than 50 years. The last plant in the United States was shut down in 1972. It would possibly still be in operation had it survived until the Arab oil embargo of 1973. Factories exist in China and South Africa. Itaconic acid is made by fermentation, and new commercial processes for lactic acid, succinic acid, *l*-2,3 butanediol, and glycerol appear likely in the relatively short term. These products have large existing or potential markets at selling prices in the range of $0.75–2.00 per pound. They are made by glucose fermentations using known species of microbes. The microbiology needed to improve these processes is microbial strain selection and modification. The strain improvements can be accomplished using traditional approaches such as random mutation or through deoxyribonucleic acid (DNA) manipulation. In either case, the required changes are "minor" in the sense that they use existing metabolic pathways. The biochemical engineering input needed for these processes centers on reactor design and separation technology. The fermentations will generally be continuous, and the separation processes must deal with relatively low concentrations of organics in water. Process costs and by-product values will be relatively important in overall economics, and good process engineering and optimization are critical. Regulatory concerns are relatively minor since the products will be sold as industrial chemicals.

Considerable work during the 1970s developed microbial or enzymatic processes for the conversion of petrochemical feedstocks to more valuable forms. Much of this was aimed at high-volume commodity chemicals, with product values little greater than their inherent energy content. Here, the excellent specificity of biochemical routes was not sufficient to overcome low reaction rates and separation problems. More recently, efforts have been directed at speciality chemicals. For example, a prototype process exists for the manufacture of primary alcohols and diols from linear aliphatics in the range C_6 to C_8. This controlled oxidation uses an extracellular enzyme in the prototypical process, and the stabilization and immobilization of that enzyme may well yield a commercially viable process for upgrading natural gas liquids. Longer-term prospects are to engineer the metabolism of organisms to use the feed hydrocarbon as their primary energy source and to optimize specific oxidations.

5. BIOCHEMICALS—SPECIALTY CHEMICALS

Biochemicals with costs low enough to be successful chemical intermediates tend to be small molecules formed by the energy-producing metabolism of a microbe. By specialty chemicals we mean molecules that are sold into end-use applications without further chemical modification. They are typically larger than the chemical intermediates, although not yet polymers, and they typically command a much higher price. Their manufacture by biochemical routes is an extremely attractive application of the emerging biotechnology. The recently announced microbial route to ascorbic acid shows the potential of this approach. Antibiotics may also fall into this category as they approach "commodity" status. Certainly, many agricultural chemicals, both existing and new, are candidates for manufacture through biotechnology.

Regulatory concerns are becoming increasingly important for agricultural chemicals. The difficulty in proving safety and efficacy of a new agricultural chemical approaches that for a new pharmaceutical. However, it does remain somewhat easier and it is a more familiar process to chemical companies. Also, the scale of manufacture is typically millions of pounds per year, which is a scale that is comforting to chemical companies. Specialty chemicals, particularly agricultural chemicals, should prove an exciting field for biochemical engineers and a profitable one for those chemical companies willing to make a long-term investment in biotechnology.

6. BIOCHEMICALS—PHARMACEUTICALS

Pharmaceuticals tend to have market values that are essentially independent of their manufacturing cost. Efficacy and safety considerations are of paramount importance, and regulatory concerns dominate the commercialization process. Raw material costs, process engineering, and by-product values are of secondary concern. Chemical engineering has played an important but relatively small role in the pharmaceutical industry compared to the role played by the chemical, biological, and medical sciences. Manufacturing processes must be frozen early in the regulatory cycle, and there are few subsequent opportunities for process innovation. This situation is unlikely to change. The advent of genetically engineered organisms will increase the scope of biochemical engineering primarily in the area of separation processes.

Antibiotics are the most common fermentation products of the pharmaceutical industry. They are typically made in batch fermentations having cycle times of 24 hr or more and giving product concentrations typically less than 1%. There has been talk of continuous antibiotic fermentations as patents have expired and the products have become "bulk" chemicals with selling prices below $10/lb. However, this faces the difficulty that the finely tuned mutants tend to revert to less productive strains when propagated through many generations.

New products currently resulting from genetic engineering are largely limited to proteins. These proteins tend to be intracellular so that cell rupture is needed to release the product. Rupture gives a complex mixture of proteins in dilute solution from which the target protein must be separated by means such as affinity chromatography, ion exchange, or ammonium sulfate precipitation. Substantial engineering would be needed to develop suitable separation process should any of the intercellular products achieve high volumes. At the moment, however, volumes remain low, selling prices are dollars or even kilodollars per gram, and what are essentially laboratory separation techniques remain suitable for "full-scale" production.

A dream of genetic engineers is to turn microbes into efficient chemical factories for a wide variety of compounds. However, relatively little progress has been made for small but complex molecules such as antibiotics. The industry has used conventional mutation and selection techniques to improve production, and further success ultimately can be expected through controlled DNA manipulation. However, success is far from automatic. By way of illustration, one large pharmaceutical company has just abandoned microbial production of an antibiotic in favor of a 20-step synthesis that gives less than 10% overall yield. Despite improvements in the microbe, the antibiotic concentration in the fermentation broth remained too low for large-scale production.

Over 2000 enzymes are known, and more than 50 new enzymes are discovered every year. Furthermore, the catalytic activity and specificity can sometimes be altered dramatically by changing the reaction environment or by replacing a key metal atom at the active site on the enzyme. Enzymes hold great promise for becoming industrially important catalysts, but this promise is so far largely unfulfilled. Major applications to date are in foods, laundry products and other cleaning applications, and the previously mentioned isomerization of glucose to fructose. Their manufacturing economics are similar to pharmaceutical proteins. It is likely that enzyme technology will benefit significantly from the current, heavy emphases on pharmaceuticals. Clearly, pharmaceuticals rather than enzymes are driving the technology.

7. BIOTECHNOLOGY COMPANIES

More than 200 new companies were formed in the United States in the late 1970s and early 1980s to exploit perceived opportunities in biotechnology. Some sources list 1500 biotechnology firms worldwide. Most biotechnology companies still operate in the red, and some bankruptcies or forced mergers have already occurred. Although infant mortality is common among start-up companies, it seems much higher in the field of biotechnology than in other high-technology fields such as electronics. Reasons for this are twofold: Many companies had unrealistic or inappropriate product goals, and many companies with excellent product goals have been undercapitalized.

Interferon was the target of far too many companies, and the product has proved rather disappointing in terms of efficacy. Even had it been more effective therapeutically, regulatory delays inherent in all pharmaceutical products would have caused major problems for all but the best-funded companies. It has been estimated that venture capital investments in biotechnology exceeded $1 billion. Although seemingly a large amount of money, the investment was highly skewed toward a few companies. The typical start-up operation had initial capital of less than $2 million. This amount is too small to bring a new pharmaceutical to market. Indeed, it usually proved too small even to develop technology of value to larger, better funded companies. Who will be doing pharmaceutically oriented biotechnology in the 1990s? Our prediction is that this work will be confined to established drug and chemical companies plus a scant few—perhaps two or three—survivors of the biotechnology start ups.

The future for small biotechnology companies is hardly more attractive when they aimed for nonpharmaceuticals. Agricultural chemicals are an extremely promising area, but the costs of bringing a new product to market are enormous. Again, established companies plus a select few new ones will survive.

Few of the new biotechnology companies were aimed at chemical intermediates. Perhaps none will survive with bulk chemicals as its major thrust. However, this fact understates the potentially large role that microbial processes will play in the manufacture of chemical intermediates. It merely restates the need for investments so large that only existing companies can afford them. Regulatory delays tend to be small compared to pharmaceuticals, but investments for capital equipment are much larger. Major product and process innovations will occur in the next decade, and some of the important technology may well be developed by entrepreneurs. However, the products will rarely be brought to market by small companies.

8. MICROBIAL PROCESSES FOR BIOCHEMICALS

This overview of biochemical engineering places substantial emphasis on the manufacture of chemical intermediates and other high-volume compounds. This is a long-term view but a suitable one for researchers just now entering the field. Processes for such compounds are usually continuous and must not involve exotic separation steps. Some ideal characteristics of a microbial process are the following:

The product should be excreted through the cell wall.

The microbe must tolerate the product at high concentrations.

The product should not be growth associated so that it is produced in the stationary or idiophase without need for comensurate increases in cell mass.

The microbe should be genetically stable and persistent in the idiophase.

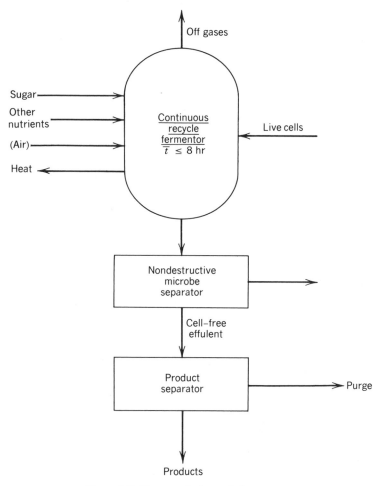

Figure 1.2 The generic fermentation process.

The microbe should function in a free aqueous suspension at high viable cell densities. It must not be too fragile.

The media cost should be low.

These criteria are met in a number of cases. They are satisfied by several bacterial genera that metabolize glucose or xylose to pyruvic acid and then generate any of ethanol, acetic acid, glycerol, lactic acid, butanediol, acetoin, or succinic acid via the Embden-Meyerhof pathway. Other metabolic paths exist that satisfy the above criteria; and in the long run, genetic engineering should be able to create microbes with new pathways leading directly to desired products.

Whenever the microbes satisfy the above criteria, the idea process will be a continuous fermentation with recycle of live microbes (see Figure 1.2).

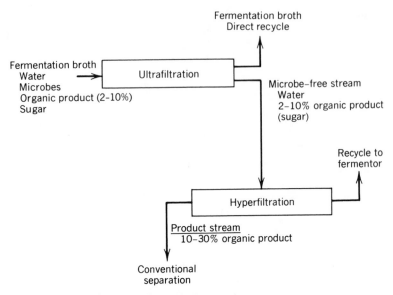

Figure 1.3 Conceptual separation processes.

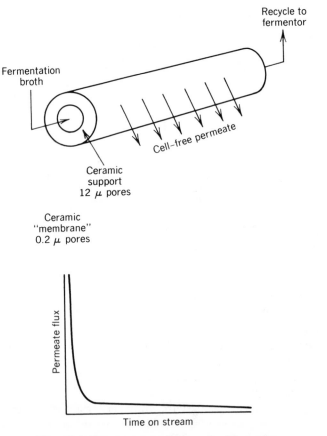

Figure 1.4 Generic process for recycle of live cells.

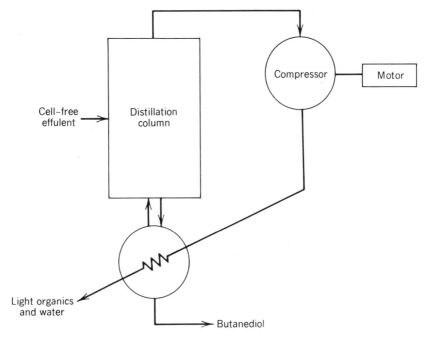

Figure 1.5 Separation scheme for butanediol process.

During start up, an innoculum is batch fermented in the usual fashion. Complex nitrogen sources (e.g., beef extract) are acceptable from a cost viewpoint since the cells, once grown, will be used and reused many times. After a desired cell destiny is achieved, nitrogen and other noncarbonaceous nutrients are eliminated (or greatly reduced) from the feed stream so that cell growth becomes minimal. The inputs to the fermentor consist of a sugar solution and possibly air. Residence times in this continuous fermentation process tend to be rather short, typically 8 hr compared to 24–48 hr for a batch fermentation. The output stream contains live cells, residual sugar, and the desired product(s). Figure 1.3 shows a conceptual separation process to be used on this effluent stream. The first step is a microfiltration to remove a cell-free product stream from the fermentation broth. Polymeric ultrafilters are the usual choice for this separation, whereas ceramic microfilters have recently shown promise. Both devices are operated in the cross-flow mode to avoid blinding the filter with cells (see Figure 1.4). The author prefers the ceramic microfilters because of their ability to withstand repeated sterilization by almost any technique. The effluent from the filter is mostly water with typically 2–10% of the desired product. If the product is a large molecule, its concentration can be enhanced by membrane separation techniques such as hyperfiltration or pervaporation. See Chapter 8. However, conventional distillation with vapor phase recompression can be a viable alternate, particularly when the product is a higher boiler than water. Figure 1.5 illustrates such a separation for the butanediol process under development at Rensselaer Polytechnic Institute. The compressor supplies the relatively small sensible heat needed to increase the temperature of the overhead stream to above that in the reboiler. The relatively large latent heat of vaporization is then recovered and is reused when the overhead stream condenses in the reboiler.

9. FUTURE FOR BIOTECHNOLOGY

Biotechnology has many possibilities. In the long run, it can revolutionalize medical, agricultural, and industrial technology. However, the time span required for this revolution is decades. By analogy, biotechnology seems roughly equivalent to the position of computer technology in the 1950s. The ultimate requirements for trained personnel are tremendous, but at the moment the need for biotechnologists and particularly for biochemical engineers is rather modest.

The major short-term growth in biotechnology is occurring in pharmaceuticals and agricultural chemicals. This book stresses those aspects of biotechnology where engineers will play the greatest role. The emphasis on separation processes is not accidental. For the longer term, the growth in biotechnology may shift toward chemical intermediates and fuels. Separation processes will remain of major importance for biochemical engineers, but there will also be greater opportunities for the design of continuous, large-scale fermentation processes.

REFERENCES

Chemical & Engineering News, **63,** p. 14, Sept. 30, 1985.

2

MICROBIAL CELLS AND ENZYMES

HENRY R. BUNGAY

Professor of Chemical and Environmental Engineering
Rensselaer Polytechnic Institute
Troy, New York

LENORE S. CLESCERI

Associate Professor of Biology
Rensselaer Polytechnic Institute
Troy, New York

GEORGE T. TSAO

Professor of Chemical Engineering and
Director of Laboratory of Renewable Resource Engineering
Purdue University
West Lafayette, Indiana

1. THE CELLULAR LEVEL

Industry puts microorganisms into unnatural environments. Substrate concentrations can be many times those that the microorganisms encounter in lakes, in streams, in soil, on plants, in animals, or in decomposing organic materials. On the other hand, microorganisms in rich industrial media grow rapidly, deplete nutrients, and create dilute systems except for relatively high concentrations of the products of their metabolism. Intense mixing and aeration in industrial processes are also unnatural and cause severe shear on cells. Many industrial strains come from molds that commonly grow on surfaces, sporulate, and travel in air currents. In an industrial fermentation, organisms exist mainly in a vegetative state surrounded by water. They respire with dissolved oxygen instead of with oxygen directly from air and do not sporulate until late in the process after nutrients are exhausted. Solid-state fermentation with growth on a solid percolated with air is much more typical of natural growth and may be more suitable for many strains.

1.1 Spatial Effects

Cultures are not composed of identical cells: There is spectrum ranging from cells that just divided and have zero age to cells that are old and dying. Furthermore, localization of cells affects their level of activity. A central cell in a chain of streptococci cannot have identical properties to an end cell that has more direct access to nutrients. A clump of staphylococci has interior cells with very restricted contact to the surrounding medium and more resistance to diffusion of nutrients and wastes. Cells attached to a solid surface must transfer materials on one side only. A mold filament has a growing tip, whereas other portions have little growth and impaired mass transfer because of steric effects and competition. Although the permutations of location, mass transport, cell age, and cell physiology defy analysis, there are implications to process engineering (Bungay et al., 1983).

Obviously, when stirring a fermentation the cells should not settle to the bottom of the reactor and compact. However, gentle mixing or bubble aeration will keep most cells pretty well suspended. Some mixing is needed to disperse bubbles and prevent large bubbles from escaping through the medium with little transfer. In a viscous broth, rather intense mixing may be required to rupture bubbles and keep them small. The limiting factor in oxygen transfer in bubble systems is the liquid film resistance. We can make liquid films thinner by mixing more intensely, but the effects on microbial cells are unclear.

Gradients of oxygen, nutrients, and product concentrations are established by microbial metabolism. As these gradients extend from the cell surface toward the bulk of the medium, there may be major inhomogenieties due to eddy diffusion and the activities of other cells. Furthermore, the location changes for suspended cells. They may be in a region rich in oxygen

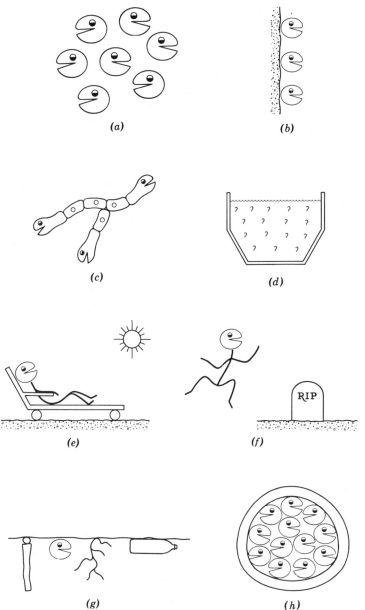

Figure 2.1. Spatial effects and mass transfer at the cellular level. (*a*) Interior cells in a clump, floc, or slime face additional mass transfer barriers; (*b*) cells at an impermeable surface can transfer only at the opposite side; (*c*) tip cells are growing and have better mass transfer than sterically hindered cells; (*d*) cells in submerged fermentation have an unnatural environment; (*e*) cells on solid nutrients approach conditions that are very common in nature; (*f*) some cells are vigorous, whereas others are dead or dying; (*g*) real interfaces collect complicated films of surface active agents, cells, and cell debris; (*h*) cells tightly packed in a bead face diffusion resistance from the walls of the bead; CO_2 evolution can rupture the bead.

near a bubble or an air–water interface and later may be swept by circulation of the fluid to a new location far from any oxygen source and thus be starved for oxygen.

Real systems have many different surface active agents that can aggregate at interfaces or on the cells. Microbial cells are also attracted to an interface. When cells, lipids, proteins, and the like, are packed in the same region, mass transfer can be severely impaired. Consumption and excretion by the cells can alter the amounts of surface active substances. The effects of mixtures of cells and surface active agents on mass transfer should be a fertile research area.

Natural attachment to a surface may be initiated because the concentration of an adsorbed nutrient may be greater than its concentration in the medium. After layers of cells coat the surface, there is no reason why the concentration of adsorbed nutrients would differ much from that for freely suspended cells. The situation is different for colonies or slimes on the benthos of a stream or lake that benefit from settling nutrient particulate matter.

It may seem wise to operate commercial processes with freely suspended cells in order to maximize surface for mass transfer, but other factors may be more important. For example, a reactor packed with a material to which cells are attached maintains a very high cell concentration. Cells immobilized in gel particles can be much more dense than cells in suspension. Cells sometimes function well when they are so closely packed, and fluid flowing past immobilized cells should flush out wastes that limit growth in batch cultures. High cell density is also achieved by separating and recycling cells from the effluent. Mass transfer on a per cell basis could be terribly inefficient for densely packed cells, but their sheer numbers lead to high overall rates. There are also cases in which activity of closely packed cells is higher than that of the same type of cells in a relatively thin suspension. For example, ethanol production by immobilized cells can be reasonably high on a per cell basis.

Mass transfer for immobilized cells can have serious problems. The essence of immobilization is to achieve high rates, but oxygen supply to a cell paste inside of beads or in the shell space of a hollow fiber reactor or to dense cell attachments to a supporting surface may be limiting. Furthermore, carbon dioxide may be evolved rapidly to burst beads or fibers or to push cell aggregates away from or through their support. Figure 2.1 illustrates some points made in this section.

1.2 Relative Motion of Cells and Medium

When cells are attached or entrapped (immobilized), the medium may flow past at a high rate and can reach the cells by diffusion and/or convection. However, suspended cells tend to have little motion relative to the medium. Flagellated cells force their way through the medium either at random or in response to tactile forces, concentration gradients, or illumination, but the velocities are miniscule.

The centrifugal forces of mixing have little effect on cells because their densities are so nearly the same as that of the medium. Even at the tips of rapidly rotating impeller blades, the centrifugal force could impart motion of only the same order as that of Brownian motion. However, there is a phenomenon that causes cells to move at rates of nearly 1 mm/sec. Fermentation broths, silt particles, fine suspensions of ion exchange resins, and the like, can be stirred in a beaker, and within seconds of removing the stirrer, a clear annulus forms concentric with the walls. This is analogous to the "tubular pinch zone" observed for particles in Poiseuille flow in a tube (Segre and Silberberg, 1961). A particle released near the wall migrates toward the center, whereas one starting at the center moves toward the wall. Cells congregate in an annulus. There the forces due to streamlines are equal to the Magnus forces caused by particle rotation. Oliver (1962) showed that a particle drilled and packed with heavier material eccentricly to prevent rotation moved to the center and formed no annulus. The migration effect increases with the cube of the particle diameter until the particle becomes so large that the velocity field is distorted. For particles less than 1 μm in diameter, the migration effect is small compared to Brownian motion. Microbial cells in the range of 2 to several hundred μm are affected strongly, and larger particles have diminishing migration effect because their size distorts the streamlines.

2. SECONDARY METABOLITE PRODUCTIVITY

Early microbial physiologists looked upon the exponential phase of microbial growth as the major and even sole elaboration of metabolic products. In the 1920s, it became apparent that older cultures of certain microorganisms have an extensive biosynthetic capability seemingly separate from energy metabolism and cellular component synthesis activities. These secondary products of metabolism are normally elaborated during the limited growth that is characteristic of the stationary phase in microbial physiology. Experience has shown that the major contributors to the vast array of commercially important secondary metabolites among the microbes are the actinomycetes and the filamentous fungi.

2.1 Metabolic Control

In recent years, the separation between trophophase (growth phase) and idiophase (secondary metabolite-producing phase) has begun to fade, and in many cases the amount of secondary metabolite produced is shown to depend not only on cell mass and physiological age but also on the excessive production of suitable precursors during ebbing primary metabolism (Drew et al. 1977). Except perhaps for a different time frame, the fundamental control processes of regulatory genes, regulation of enzyme action, and regulation of enzyme synthesis apply to secondary as well as primary metabolism.

With each newly discovered secondary metabolite, considerations of yield, stability, and toxicity immediately arise. From a commercial point of view, the product must qualify on all three points, but we have most control over the yield.

There are many examples of the effect of low molecular weight metabolites on a yield of secondary metabolic pathways (Martin et al., 1980). Usually, the mode of action of the effect is not clear and may be either at the level of enzyme synthesis or the enzyme action. It is clear, however, that there is an inverse relationship between growth rate and product yield. Readily available sources of carbon and nitrogen tend to suppress the production of secondary metabolites. During rapid growth, it makes biological sense for organisms to suppress the elaboration of substances, such as antibiotics, that may be detrimental to their own well-being (Demain, 1974). In mixed culture studies we have observed distinctive differences in microbial floc morphology as a function of metabolic availability of carbon source. The observed floc in the readily available carbon source system rapidly developed a filamentous component that was suppressed by media without low molecular weight metabolites. This could be related to competitive advantages produced by antibiotic elaboration in the suppressed system, illustrating the role of carbon availability.

Frequently, it is found that during the screening of nitrogen sources for secondary metabolite production, a slowly metabolized amino acid produces the highest level of product. Conversely, ammonia and other rapidly utilized nitrogen sources have been found to suppress product formation. Streptomyocin, candicin, chloramphenicol, rifamycin, penicillin, and cepholosporins are among those secondary metabolites that can be regulated by nitrogen metabolism (Demain, 1982).

2.2 Strain Development

Most microbial products of commercial interest come from multienzyme reactions, that is, they are the products of complex biochemical pathways that bear little resemblance to primary metabolic pathways. Whereas protein products that are the product of a single gene are prime candidates for yield enhancement via molecular cloning techniques, multienzyme products will continue to require other approaches toward improvement in productivity because of the genetic complexity of their biosynthetic pathways. Enhanced productivity of multienzyme products such as antibiotics, amino acids, and vitamins by gene cloning has proven to be much more complex and difficult. The classical approach to yield improvement of secondary metabolites has been random mutagenesis of a producer, with screening of survivors for enhanced productivity. The process is empirical, and many of the hyper-producing strains must subsequently be nurtured by cultural manipulations to obtain satisfactory growth conditions for large-scale production. Therefore, the commercial strains become fairly distant relatives of their wild ancestors. This process for yield enhancement has been remarkably successful for many secondary metabolites (Elander, 1982; Fiedler et al., 1981). However, the process has not supplied the biochemical reasons for increased productivity. In order to use gene manipulation for multienzyme products, detailed knowledge of pathway enzymology and pathway regulation is needed. This is rarely known (Hutter et al., 1978), and until this is available, productivity enhancement will continue to require the classical approach supplemented perhaps with some alternate directions such as directed biosynthesis and mutasynthesis.

2.3 Alternative Approaches

Among the tools that the biotechnologist has for the optimization of productivity in the fermentation industry is the deregulation of certain biochemical pathways to effect the accumulation of suitable secondary metabolite precursors. This approach requires a sophisticated knowledge of the biosynthetic pathways of the products. In some instances, the approach of directed biosynthesis has been found to be successful in improving the yield of secondary microbial products. In directed biosynthesis, select additives to a fermentation broth direct the biosynthesis of a particular product by acting as a positive effector or stimulator of a key stop.

The earliest example of this approach was the inclusion of phenylacetic acid or a close chemical relative in media used for production of benzyl penicillin. In this case, phenylacetate proved to be a direct precursor incorporated into the penicillin. Other examples include the use of chloride in chlortetracycline synthesis, sarcosine in production of actinomycin, and cobalt in vitamin B_{12} synthesis (Rose, 1979).

A recent example of directed biosynthesis is illustrated by the effect of feeding branched chain L-amino acids to a nikkomycin fermentation (Fiedler et al., 1981). The production of these chitin synthase inhibitors was stimulated up to 5.8 times depending on the particular amino acid used. Whether the amino acids have been directly incorporated into the nikkomycins has not been established, but the concept of directed biosynthesis does not necessarily include incorporation. Where detailed biosynthesis details are lacking, directed biosynthesis offers an intermediate approach in the fermentation optimization process. It is one that can inexpensively and sometimes dramatically affect the productivity of a fermentation.

The process of mutasynthesis has been used with some success to bypass the cell's precursor machinery through specific mutation of a gene that codes for one precursor of the organism's natural secondary metabolite (Daum and Lemke, 1979). This leads to the synthesis of an incomplete secondary me-

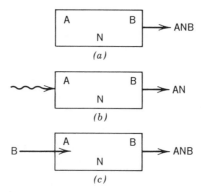

Figure 2.2 The process of mutasynthesis. (*a*) Cells synthesize antibiotic; (*b*) incomplete antibiotic from mutated cell; (*c*) addition of permeable precursor to medium results in production of antibiotic.

tabolite, which is subsequently completed by the addition of precursor to the medium (Fig. 2.2). This can get around the problem of feedback inhibition if the organism is permeable to the precursor.

Another way to avoid the feedback problem is to use resistant mutants that show decreased sensitivity to accumulation of product on their productivity. In the case of cephalosporin C, mutants resistant to valine analogues yielded a higher frequency of superior producers during the screening of mutagenized cells (Elander, 1982). The resistant mutant approach has been very successful in amino acid overproduction as well as in certain secondary metabolite fermentations.

A completely different direction that can be taken in enhancing productivity is one that has been mentioned earlier and involves the use of recycling cells or sustaining high concentrations of active cells in reactors. Cell, enzyme, and cofactor immobilization on inert supports can provide for recycling in continuous flow systems. Alternatively, high solids can be sustained by ultrafiltration as discussed in Chapter 3.

In situations in which solid substrates are used, the substrate itself may serve as the immobilizer. For example, in batch cellulolysis systems using microbial enzymes, recycling of cellulase enzymes (except for β-glucosidase) can be effected by pH adjustment of the spent wood to release active enzyme for recycling into another batch process (Sinitsyn et al., 1983).

Systems that depend upon the sorptive role of microbial cells (as in waste treatment) use high solids systems achieved through various recycling schemes. These high concentration mixed culture systems are capable of diverse metabolic activities and, very likely, considerable secondary productivity under the highly competitive existence encountered in waste treatment.

3. MIXED CULTURES

An enginering view of mixed cultures was given by Haas et al. (1981). Microbial interactions are also discussed in books by Wicklow and Carroll (1981), Bushell and Slater (1981), Bazin (1982), Bull and Slater (1982), and Klug and Reddy (1984).

Resolution and enumeration of population components can be difficult, so laboratory studies often use organisms of different sizes to permit electronic particle counting. Sometimes morphological differences are sufficient to permit resolution with microscopic examination of a counting chamber. Tedious, less precise differential counting can be performed by plating on solid media. When colonial differences are slight, selective plating media must be used. Indirect methods are analyzing pigments assumed to be associated with particular organisms and assaying other biochemicals of a general nature such as adenosine triphosphate (ATP), deoxyribonucleic acid (DNA), muramic acid, or biochemicals produced by a given species.

Karel et al. (1985) present a comprehensive review of the literature on the physiology and mass transfer limitations of immobilized whole cells. They determine by use of radioisotope labeling the special distribution of microorganism growth rate in hollow fiber bioreactors. Another technique for probing cell metabolism is flow cytometry (Hofmann and Sernetz, 1983; Horan and Wheelers, 1977). The method can analyze individual cells passing through a laser beam; concentrations of internal biochemicals are estimated from the cell's fluorescence. This technique can also be used to resolve and count mixed populations and to follow biochemical changes in the systems. Various methods for studying mixed cultures are reviewed by Lovelock and Davies (1978). Recent attempts have been made to study the growth cycle of hybridomas in a spinner flask to determine when maximum production of monoclonal antibodies are formed (Altshuler, G. A., private communication, 1985).

3.1 Nomenclature

Some common terms for microbial interaction are

Neutralism: lack of interaction

Commensalism: one member benefits while the other is unaffected

Mutualism: each member benefits from the other

Competition: a race for nutrients and space

Amensalism: one adversely changes the environment for the other

Parasitism: one organism steals from another

Predation: one organism ingests another

Synergism: cooperative metabolism to produce a substance not produced by either alone

A more precise system has been developed by Fredrickson (1977). It is virtually impossible to have a truly single-interaction system because organisms occupy space and affect the chemical and physical aspects of their environment in ways that may stimulate or inhibit other types of organisms present.

3.2 Growth Limitations

Stoichiometric limitation of growth is well understood, but limitation by rates of biochemical processes has been virtually ignored. Carbon, hydrogen, nitrogen, sulfur, phosphorus, and other elements supplied to the culture become cell mass or are consumed by metabolic processes. One constituent in lower proportion than the others can limit growth. Mixed cultures are complicated because nutritional requirements differ and some organisms excrete nutrients for others.

If all constituents of the medium are in surplus (and none is toxic), there is still a maximum growth rate that must relate to the kinetics of key biochemical steps. Some rate processes, such as transport through a membrane, synthesis of a macromolecule, and diffusion of an intermediate from one enzyme to another, are going as fast as they can, and no alteration of the composition of the medium produces a higher rate for this step.

In continuous culture, generally a growth-limiting nutrient operates stoichiometrically. Typical dilution rates for chemostats are well below the maximum specific growth rates of the organisms, thus kinetic limitations do not come into play. Several studies in continuous culture with control of substrate concentration or in chemostats with very high dilution rates have demonstrated that bacteria can double in 4–8 min (Gitelson et al., 1974). This is definitely approaching maximum growth rate, and kinetic limitations must be important. In fact, an ingredient that was growth limiting because of its stoichiometry at lower dilution rates may become nonlimiting at very high dilution rates.

Although the concept of a single limiting substrate has been very useful, real life is seldom that simple. The excellent agreement of real data with the Monod equation probably depends on having a fairly large excess of all but one ingredient. When proportions of ingredients of the medium are closer and shifting because of population changes in mixed cultures, there are very likely switches in limiting ingredients for several types of organisms. Further research to unravel the mysteries of growth rate dependencies would be highly desirable.

The Monod equation is not time dependent and is valid only for steady state. Dynamic models should consider lags in adjustment of μ, the specific growth rate coefficient, in response to changes in substrate concentration, but most models of mixed culture dynamics ignore these time constants and assume that μ adjusts instantaneously. Mateles et al. (1965) showed that some adjustment is practically instantaneous if the step increase in substrate concentration is small, but several hours may be required to reach the final value of μ for larger deviations.

3.3 Elective Cultures

Elective cultures accommodate well to open systems and participate in creation of their environment. Natural populations can undergo remarkably severe changes. Some examples are the rapid shift from one predominant algal species to another in a lake or pond and sudden shifts in the bacterial composition of a stream. Cassel et al. (1966) using pigments as indexes of bacterial concentrations in an activated sludge unit observed pronounced changes. A species that was dominant for several days fell to undetectable levels and might reappear much later as a major component of the mixed culture. Such behavior makes it highly tenuous that studies of defined systems with a few components in the laboratory will be acceptable models of natural systems with many biological components.

A qualitative explanation of erratic fluctuations in natural populations can be based on the oscillations of simple systems. In nature it is highly likely that a given organism is participating in many distinct microbial interactions, some of which are oscillatory. Furthermore, these interactions would be expected to be asynchronous. The oscillations in population of a microbial species could reach high numbers when the peaks of several distinct asynchronous interactions are reinforcing and could fall to very low numbers when the valleys of the distinct interactions overlap.

Populations change in response to stress. In most natural ecosystems not already subject to great stress, multitudes of different species are present. A few may constitute high proportions of the total numbers present, whereas very few individuals may represent the least abundant species. Spores or dormant cells may initiate new growth should conditions become favorable. Airborne particles can constantly reinoculate an aqueous system. Applying a stress such as a higher temperature may favor a species that was already active, a dormant species, a variant strain, or a naturally occurring mutant. Only very heavily stressed environments such as a hot stream with high concentrations of hydrogen sulfide develop essentially pure cultures. Weaker stresses tend to lead to new dynamic population balances, although it is usual to find markedly reduced diversity in a stressed ecosystem compared to its condition before the stress was applied.

The ability to accommodate to stress is a selective advantage for natural populations; life persists despite changes in pH, temperature, redox potential, and nutrients. Were there not organisms present to metabolize toxic wastes such as phenol, undesired substances would accumulate. Responses to stress are usually slow; thus shocking a waste treatment plant with phenol can be disastrous, whereas gradual addition allows those organisms metabolizing the toxic substance to achieve significant numbers. This draws attention to another point: the presence of a few individuals may not be sufficient to allow recovery from stress if the entire population is nearly wiped out. Whereas gradual stress may permit beneficial associations for the needed strains, violent stress may so severely inhibit the associated strains that the desired strains are also lost. Furthermore, gradual stress is accompanied by metabolism of the toxic material so that highly inhibitory concentrations are not reached. Abrupt shocks may reach concentrations that inhibit even those organisms that are able to use the material.

The persistence of diversity is not seen with laboratory systems. For example, in a chemostat with no particular stresses applied, mixed populations lose heterogeneity. Direct competition leads to take over by faster organisms while others washout. In a natural system with flow, a very few well-acclimated species should win out while most others wash away. In fact, several types tend to be abundant, others are more scarce but significant, and many others persist indefinitely at low numbers. However, natural systems are not uniform, and direct extrapolation from well-mixed laboratory reactors is not valid.

A powerful competitive stress has been applied to continuous cultures by use of feedback control of substrate concentration (Bungay et al., 1981). Selection of faster-growing organisms depletes the substrate and causes the controller to increase feed rate, and thus residence time in the vessel is reduced. An elective culture with many constituents is quickly converted to a population of just one or two different organisms.

The endlessly complicated variations, evolution, and responses of natural ecosystems would seem to defy rational analysis. Laboratory studies of defined mixed cultures are entertaining and of definite value as a foundation for commercial mixed culture fermentations, which are also artificial systems. However, extrapolation of defined systems to natural systems is highly dubious in light of the complexities that have been discussed. Explanations for the behavior of mixed cultures in systems such as those for biological waste treatment will require more than laboratory experiments with defined cultures.

3.4 Competition

The theory of simple competition in a well-mixed continuous culture is straightforward and is developed in reviews by Harder et al. (1977) and Harder (1973). Wilder et al. (1980) devised various control algorithms that will maintian both competitors when substrate concentration determines which grows faster. Stephanopoulous et al. (1979) considered the same problem and also developed control systems to ensure coexistence. This work is important because in special cases a synergistic mixed population might be maintained in a commercial mixed culture process by a feedback control system.

3.5 Predation

Tsuchiya et al. (1972) found typical sustained oscillations followed by decaying oscillations to apparent steady states that could be followed by resumption of oscillations. These observations were confirmed by Dent et al. (1976) who also used *Dictyostelium discoideum* and *Escherichia coli*. When a temporary steady state was established, a change in an experimental variable produced a different steady state. Bazin (1978) and Bazin et al. (1978) studied this same system with emphasis on uptake of nutrients from the medium.

Drake and Tsuchiya (1977) operated a two-state continuous culture with *E. coli* in the first stage feeding a second stage that had *Colpoda steinii* as well. The system tended to achieve steady state.

3.6 Other Associations

Oberman and Libudzis (1973) tested mixtures of *Lactobacillus casei* and a strain of *Streptococcus lactis* that produced the antibiotic nisin. Both organisms grew in batch culture when started in certain proportions. Similar proportions led to stable mixed continuous cultures, and nisin was not detected. This association illustrates amensalism, which is overcome by adjusting relative numbers of organisms.

Meers and Tempest (1968) studied binary cultures in chemostats limited by the concentration of magnesium ion. Each culture excreted growth-promoting substances that were very effective for the producing culture and of lesser benefit to the other culture. Therefore, this association was not a simple competition but a mutualistic situation over a range of concentrations.

There are some extremely interesting associations where one organism lives within the cells of another. One example with great practical potential is the water fern Azolla that harbors algae that harness sunlight for the production of biochemicals, including nitrogen fixed from the atmosphere. This fern is one of the most efficient fixers of nitrogen, with rates much better than those of alfalfa and other legumes and many times the rate of free cyanobacteria (formerly called blue-green algae).

3.7 Biological Approaches to Population Control

Venosa (1975) has attempted control of undesirable *Sphaerotilus natans* by providing parasitic strains of *Bdellovibrio bacteriovorus*. *Sphaerotilus* is a filamentous organism often associated with the bulking of activated sludge or objectionable slimes in piping or in natural waters. The *Bdellovibrio* attacked individual cells of *Sphaerotilis* but were unable to penetrate the sheaths when long filaments had formed. It is unlikely that this control method would be effective after *Sphaerotilus* was established, but individual cells migrating to initiate new filaments might be suppressed.

Selection in continuous cultures can have a time span of a few days. When virulent phage are inoculated into a bacteria culture, oscillations of a prey–predator type begin at once but soon become damped and distorted (Horne, 1970; Paynter and Bungay, 1969). Varieties or mutants of the parasites have the ability to infect the bacterial strains resistant to the original, predominant parasites. Thus a natural population must be in an evolving complicated state of selection and mutation where the hosts develop resistant forms but new forms of the parasites attack them.

Few commercial fermentations use mixed cultures, but they remain of prime importance for waste treatment despite advances in alternative physiochemical processes. Much research is needed to gain adequate understanding of microbial interactions that determine performance and stability of processes based on elective cultures. Industrial processes based on defined mixed cultures should have increasing importance, and controlling mixed populations may be more effective than using genetic engineering to incorporate all the traits in one organism.

4. IMMOBILIZED CELLS AND ENZYMES

Enzymes are superb catalysts because of high specificity and high turnover rates. With few exceptions, all reactions in biological systems are catalyzed by enzymes, and each enzyme usually catalyzes only one reaction. Although amino acid sequences are known for several enzymes and other proteins, lack of a molecular formula for most enzymes means that concentration cannot be expressed in molarity. An international unit (IU) of an enzyme is defined as the amount that produces 1 μ mol of its reaction product in 1 min under its optimal (or some defined) reaction conditions. Specific activity (IU per unit weight) is an index of enzyme purity.

Considering each enzyme molecule to have an active site that must first encounter the substrate (the reagent) to form a complex so that the enzyme can function, the reaction scheme is

$$\text{Enzyme + Substrate} \underset{2}{\overset{1}{\rightleftharpoons}} \underset{3}{\text{Complex} \rightarrow} \text{Product + Enzyme}$$

These reactions are assumed to be in equilibrium soon after the enzyme is exposed to its substrate. The rate of the product formation thus depends upon the concentration of the enzyme–substrate complex for the equation

$$k_2[ES] = k_1[E][S] = k_1[S]([E]_T - [ES]) \tag{2.1}$$

and

$$V = \frac{d[P]}{dt} = kf\,[ES] \tag{2.2}$$

where

$$
\begin{aligned}
[E]_T &= [E] + [ES] = \text{total enzyme concentration}\\
[S] &= \text{substrate concentration}\\
[ES] &= \text{concentration of enzyme--substrate complex}\\
[P] &= \text{product concentration}\\
t &= \text{time}\\
k_1, k_2, k_3 &= \text{kinetic constants}
\end{aligned}
$$

This leads to

$$V = \frac{V_{max}\,[S]}{K_m + [S]} \tag{2.3}$$

where $K_m = k_2/k_1 =$ Michaelis constant.

It is difficult to estimate the coefficients from a curving plot of rate versus substrate concentration, but the relationship is linearized by taking reciprocals of both sides:

$$\frac{1}{V} = \frac{K_m\,V_{max}}{S} + \frac{1}{V_{max}} \tag{2.4}$$

A linear plot of $1/V$ versus $1/S$, (the Lineweaver-Burk plot) indicates K_{max} from experimental data.

Kinetic behavior becomes complicated when there are two chemical species that both complex with the enzyme molecules. One of the species might inhibit the enzyme reaction. Depending upon the nature of the complex, different inhibition patterns will yield different kinetic equations.

5. ENZYME IMMOBILIZATION

The development of wide industrial application of enzymes is dogged by high cost, so it is attractive to immobilize enzyme molecules for repeated use. Methods for immobilization are adsorption, covalent bonding, and entrapment. Semipermeable membranes in the form of sheets or hollow fibers can retain the enzyme and at the same time allow smaller molecules to pass. Polyacrylamide gel, silica gel, and collagen have been used for entrapment of biologically active materials such as enzymes. Encapsulation captures enzymes by coating liquid droplets with some semipermeable materials formed in situ. Generally speaking, entrapment does not involve a reaction, and the enzyme molecules are not altered. Physical adsorption on active carbon particles and ionic adsorption on ion exchange resins are important for enzyme immobilization. A method with a myriad of possible variations is covalent bonding of the enzyme to a selected carrier. Materials such as glass particles, cellulose, silica, and even metals have been used as carriers for immobilization. Enzymes immobilized by entrapment and adsorption may be subject to loss due to leakage and desorption. On the other hand, the chemical treatment to form the covalent bond between an enzyme and its carrier may permanently damage some enzyme molecules. Immobilization yield can be used to describe the percentage of enzyme activity that is immobilized:

$$\% \text{ yield } = 100 \times \frac{\text{Apparent activity}}{\text{Starting activity}}$$

When an enzyme molecule is attached to a carrier, its active site might be sterically blocked so that its activity becomes unobservable (inactivated).

One of the most important parameters of an immobilized-carrier complex is stability of its activity. Catalytic activity of the complex diminishes with time because of leakage, desorption, deactivation, and the like. The half-life of the complex is often used to describe the activity stability. Even though there may be frequent exceptions, linear decay is often assumed in treating the kinetics of activity decay of an immobilized complex.

Wandery and Flashel (1979) compared the technical and economic aspects of carrier-fixed enzymes and soluble enzymes in a membrane reactor. Their system was acylase catalysis of n-acetyl-D,L-methionine to produce L-methionine. The membrane reactor had better economics for these conditions and assumptions.

Immobilization by adsorption or by covalent bonding often helps stabilize the molecular configurations of an enzyme against alterations such as thermal deactivation. Immobilized enzymes tend to be less sensitive to pH changes than are free enzymes. Most carriers are designed to have high porosity and large internal surface areas so that a relatively large amount of enzymes can be immobilized onto a given volume or given weight of the carrier. Therefore, in an immobilized enzyme-carrier complex, the enzyme molecules are subject to the effect of the microenvironment in the pores of the complex. Surface charges and other microenvironmental effects can create a shift up or down of optimal pH of the enzyme activity.

An immobilized enzyme-carrier complex is a special case that can use the methodology developed for evaluation of a heterogeneous catalytic system. The enzyme complex also has external diffusional effects, pore diffusional effects, and an effectiveness factor. When carried out in aqueous solutions, heat transfer is usually good, and it is safe to assume that isothermal conditions prevail for an immobilized enzyme complex. The Michaelis-Menten equation and other similar nonlinear expressions are useful with immobilized enzymes.

Immobilization of microbial cells has the same rational as immobilization of enzymes, but more functionality can be realized. For example, in order to catalyze an oxidation–reduction reaction, there must be enzymes and their cofactors. Although it is possible to immobilize an enzyme and a cofactor in close proximity so that they can interact, the technology is far more complicated than just tethering an enzyme. On the other hand, an immobilized cell contains many enzymes and cofactors. When a substrate molecule enters the cell, it can find all the necessary factors for reaction. Ultimately, the kinetics may be better for a cell-free system that has no mass transfer limitations due to cell walls and cell membranes, but multistep reactions and those requiring cofactors are presently suited to immobilized cells rather than to in vitro systems.

In the United States and elsewhere during the 1970s, there was emphasis on methodology for immobilization. Some excellent techniques were developed, and modeling and mathematical analysis covered most aspects of reactors for using these catalysts. There is concensus that the world needs more applications of cell and enzyme technology and that further improvements in immobilization or reactor design deserve low priority. To state matters differently, there is a host of capability and technology and years of experience with heterogeneous reactions that can be applied to using immobilized cells or enzymes, but very few new industrial applications seem to be forthcoming.

There are some unsolved problems. Probably stability is of greatest concern. In the last 10 years, understanding of enzyme stability has advanced very little. Another restraint to applying enzyme technology is cost of cofactors. At several hundred dollars per kilogram, cofactor cost may cripple process economics, particularly in view of losses due to instability.

Although chemical immobilization of cells is quite feasible, entrapment or adsorption have more attractive costs. Simple adsorption to carbon is a commercial method for using glucose isomerase because this enzyme is relatively stable and remains active when cells are heated and attached to carbon. This is not the usual type of cell immobilization because the cells are essentially dead. Cells inside a gel or in the shell space of a hollow fiber reactor may be very densely packed and not growing because the medium is deficient in nitrogen or some other factor. Despite mass transfer limitations due to diffusion through the gel, encapsulated cells tend to carry out rapid conversions. For example, yeast cells in gels can produce about 10% ethanol from glucose in one pass with a detention time or roughly 1 hr. In some-cases, evolution of carbon dioxide has caused rupture of the membrane, allowing cells to be released.

Flocculated cells can also pack densely but not so tightly as cells squeezed into a gel capsule. Nevertheless, rates are excellent and costs are low. The Purdue group has a very simple method for finding flocculating strains of *Schizosaccharomyces pombe*. A strain from culture collections had dispersed growth, but a few flocs form. After 12–24 hr of growth in shake flasks, the culture is allowed to stand for 10–20 min, and 90% of the fluid is siphoned from the top. Fresh medium is added, and the process is repeated many times. With removal of most of the suspended cells and retention of the flocculating cells, continual enrichment yields a culture that settles very rapidly. Such organisms are quite suitable for packed cell reactors. Although this technique is excellent for organisms that have even a small tendency to flocculate, some organisms seem to form absolutely no flocs. There may be a genetic hurdle against flocculation, and recombinant DNA approaches may be worthwhile.

Flocculated cells tend to form soft particles that may compress in a column. Careful hydrodynamic balance may provide satisfactory fluidization, but properties can change with time. Gel beads have much greater strength and permit a wider range of flow.

There is considerable confusion about the effectiveness factor for immobilized cells. If it is defined as the ratio of the reaction rate of packed cells or enzymes to the reaction rate for dispersed cells or enzymes, the effectiveness factor should be less than 1 because of diffusional limitations. In the case of immobilized enzymes, effectiveness factors are good and may approach 1 because much of the activity is on the surface of support materials and diffusion is not a serious problem. However, some packed cell reactors have effectiveness factors as high as 5. This cannot be explained by conventional chemical engineering. The mystery is caused by the way we define the effectiveness factor, because the packed cells differ morphologically and probably physiologically from freely suspended cells. They tend to be longer and of smaller diameter when packed. The concept of effectiveness factor must be refined or replaced with a more useful tool for packed bioreactors, and the biochemical engineers who undertake this task must consider some complicated microbiology and biochemistry.

REFERENCES

Bazin, M. J. *Ann. Appl. Biol.* **89**, 159, 1978.

Bazin, J. J., Sawndeys, P. T., Owen, B. A., and Kilpatrick, D. In Loutit, M. W., Miles, J. A. R. (eds.), *Microbial Ecology,* West Berlin, Springer, 1978.

Bazin, M. J. *Microbial Population Dynamics* Boca Raton, Florida, CRC Press, 1982.

Bull, A. T., and Slater, J. H. (eds.). *Microbial Interactions and Communities,* New York, Academic Press, 1982.

Bungay, H. R., Clesceri, L. S., and Andrianas, N. A. In Moo-Young, M., Vezina, C., and Robinson, C. R. (eds.), *Advances in Biotechnology.* Elmsford, NY, Pergamon, 1981, pp. 235–241.

Bungay, H. R., Bungay, M. L., and Haas, C. N. *Ann. Rep. Ferm. Proc.* **6**, 149, 1983.

Bushell, M. E., and Slater, J. H. (eds.). *Mixed Culture Fermentations.* New York, Academic Press, 1981.

Cassell, E. A., Sulzer, F. T., and Lamb, J. C., III. *J. Water Pollut. Control Fed.* **38**, 1398, 1966.

Daum, S. J., and Lemke, J. R. *Ann. Rev. Microbiol.* **33**, 241, 1979.

Demain, A. L. *Ann. N.Y. Acad. Sci.* **235**, 601, 1974.

Demain, A. L. In Krumphanze, V., Sikyta, B., and Vanek, Z. (eds.), *Overproduction of Microbial Products,* New York, Academic Press, 1982, pp. 3–20.

Dent, V. E., Bazin, M. J., In Droop, M. R., and Jannasch, H. W. (eds.), *Advances in Aquatic Microbiology,* New York, Academic Press, 1976, Vol. 1, p. 115.

Drake, J. F., and Tsuchiya, H. M. *Appl. Environ. Microbiol.* **34**, 18, 1977.

Drew, S. W., and Demain, A. L. *Ann. Rev. Microbiol.* **31**, 343, 1977.

Elander, R. P. In Krumphanze, V. Sikyta, B., and Vanek, Z. (eds.), *Overproduction of Microbial Products,* New York, Academic Press, 1982, pp. 352–369.

Fiedler, H. P., Kurth, R., Langharig, J., Delzer, J., and Zahner, H. J. *Chem. Technol. Biotechnol.* **32**, 271–280, 1981.

Fredrickson, A. G. *Ann. Rev. Microbiol.* **31**, 63, 1977.

Gitelson, I. I., Kuznetsov, A. M., Rodicheva, E. K., Fish, A. M., Chumakova, R. I., and Shcherbakova, G. Ya. *Biotechnol. Bioeng. Symp.* **4**, 857, 1974.

Haas, C. N., Bungay, H. R., and Bungay, M. L. *Ann. Rep. Ferm. Proc.* **4**, 1, 1981.

Harder, W. *J. Appl. Chem. Biotechnol.* **23**, 707, 1973.

Harder, W., Kupnen, J. G., and Martin, A. *J. Appl. Bact.* **43**, 1, 1977.

Hofmann, J., and Sernetz, M. *Trans. Anal. Chem.* **2**, 172, 1983.

Horan, P. K., and Wheeless, L. L. *Science* **198**, 149, 1977.

Horne, M. *Science,* **168**, 992, 1970.

Hutter, R., Leisinger, T., Nuesch, J. and Wehrli, W. *Antibiotics and Other Secondary Metabolites—Biosynthesis and Production,* London, Academic Press, 1978.

Karel, S. F., Libicki, S. B., and Robertson, C. R. *Chem. Engr. Sci.* **40**, 1321, 1985.

Klug, M. J., and Reddy, C. A. (eds.), *Current perspectives in microbial ecology, Am. Soc. Microbiol.,* 1984.

Lovelock, D. W., and Davies, R. *Techniques for the Study of Mixed Populations,* London, New York, Academic Press, 1978.

Martin, J. F., and Demain, A. L. *Microbiol. Rev.* **44**, 230–251, 1980.

Mateles, R. I., Ryu, D. Y., and Yasuda, T. *Nature,* **208**, 263, 1965.

Matteo, C. C., Cooney C. L., and Demain, A. L. *J. Gen. Microbiol.* **96**, 415, 1976.

Meers, J. L., and Tempest, D. W. *J. Gen. Microbiol.* **52**, 309, 1968.

Oberman, H., and Libudzis, Z. *Acta Microbiol. Pol. Ser. B. Microbiol. Appl.* **5**, 151, 1973.

Oliver, D. R. *Nature* **194**, 1269, 1962.

Paynter, M. J. B., and Bungay, H. R. In Perlman, D. (ed.), *Fermentation Advances,* New York, Academic Press, 1969.

Rose, A. H. In Rose, A. H. (ed.), *Secondary Products of Metabolism,* New York, Academic Press, pp. 2–33, 1979.

Segre, G., and Silberberg, A. *Nature,* **189**, 209, 1961.

Sinitsyn, A. P., Bungay, M. L., Clesceri, L. S., and Bungay, H. R. *Appl. Biochem. Biotech.* **8**, 25, 1983.

Stephanopoulos, G., Fredrickson, A. G., and Aris, R. *AIChE J.* **25**, 863, 1979.

Tsuchiya, H. M., Drake, J. F., Jost, J. L., and Fredrickson, A. G. *J. Bacteriol.* **110**, 1147, 1972.

Wandry, C., and Flaschel, E. In Ghose, T. K., Fiechter, A., and Blakebrough, N. (eds.), *Advances in Biochemical Engineering,* Vol 12, Berlin, Springer-Verlag, pp. 147–218.

Wicklow, D. T., and Carroll, G. C. *The Fungal Community: Its Organization and Role in the Ecosystem,* New York, Marcel Dekker, 1981.

Wilder, C. T., Cadman, T. W., and Hatch, R. T. *Biotechnol. Bioeng.* **22**, 89, 1980.

3

ANALYSIS OF BIOLOGICAL REACTORS

CLEMENT KLEINSTREUER

Associate Professor of Mechanical and Aerospace Engineering
Member of the Biotechnology Group
North Carolina State University
Raleigh, North Carolina

1. INTRODUCTION

For many years, the stirred tank batch fermenter has been the mainstay for industry, and controlled feeding of nutrients has become a standard technique. More recent advances include loop and airlift fermenters, packed or fluidized bed reactors using biocatalysts, and hollow fiber and other membrane arrangements for confining enzymes or whole cells (Schügerl, 1981; Moo-Young and Blanch, 1981; Schügerl, 1982; Michaels, 1982; Sittig, 1983; Bungay, 1984; Kleinstreuer and Poweigha, 1984). Schematics of these bioreactors are given in Figure 3.1. The airlift fermenter uses air to agitate the fluid by means of draft tubes or external recycle so that expensive mechanical mixers can be avoided. Unfortunately, airlift works poorly for many important fermentations that use viscous mold cultures that impair oxygen transfer. One very good application for airlift fermentation is the production of single-cell protein with bacterial cultures that are not very viscous.

The deep shaft fermenter is a long, relatively thin vessel that loops at a distance far below the feed point for air. Air is compressed as the medium

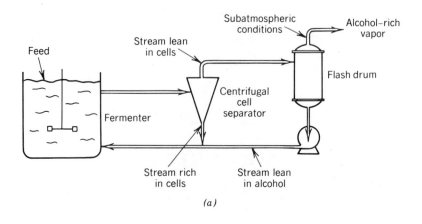

(a)

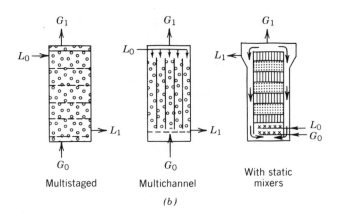

Multistaged   Multichannel   With static
                                            mixers

(b)

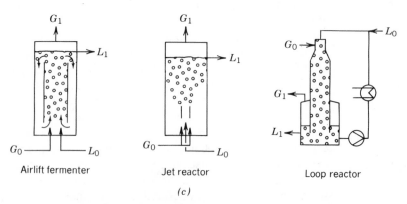

Airlift fermenter   Jet reactor   Loop reactor

(c)

Figure 3.1. Schematics of selected bioreactors. (a) Stirred tank reactor with cell recycle and product separation; (b) bubble column reactors; (c) Airlift and loop fermenters;

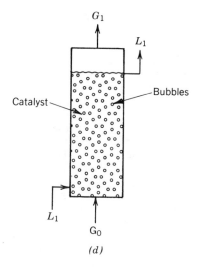

G_1

L_1

Catalyst

Bubbles

L_1

G_0

(d)

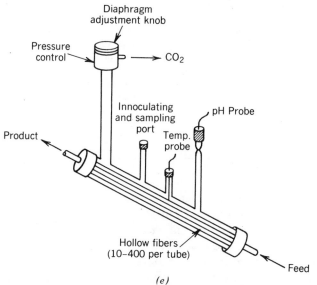

Diaphragm
adjustment knob

Pressure
control

CO_2

Innoculating
and sampling
port

pH Probe

Temp.
probe

Product

Hollow fibers
(10–400 per tube)

Feed

(e)

Figure 3.1 *(continued)*. *(d)* three-phase fluidized bed reactor; *(e)* Hollow fiber fermentor with monitoring probes and pressure control.

TABLE 3.1. Evaluation of Selected Biochemical Research Reactors

Continuous flow Bioreactors	Principal Advantages	Major Disadvantages	Basic Research Needs
Continuous flow stirred tanks w/re-cycle (Hamer, 1982)	Based on known simple technology; tanks in series	Energy intensive; moving parts; re-cycle problems; uniform mass transfer problems	Stability analysis; optimization and control
Loop and airlift fer-menters (Schügerl, 1982)	High oxygen trans-fer; no moving parts; high volume production	Energy intensive; for limited microbial systems only; uni-form mass transfer problems	Investigation of complex transfer and conversion processes; non-Newtonian flow
Fluidized/packed bed bioreactors (Andrew, 1982)	High, continuous throughput; im-mobilization of mi-croorganisms	High-pressure drop; potential overflow diffusion limita-tions; separation problems; expen-sive	Solutions to complex three-phase flow and mass transfer problem
Draft tube gas–liq-uid–solid fluidized bed (Fan et al., 1984)	High gas–liquid mass transfer; more flexible than conventional three-phase re-actors	Energy intensive with considerable pressure drop; dif-fusion and separa-tion problems in biological phase	Further testing of the draft tube fluidized bed as a bioreactor; solu-tion to three-phase flow rheology and mass transfer
Hollow fiber bio-chem. reactors (Kleinstreuer and Agarwal, 1986)	Separation of cells from feed; no moving parts	Diffusion limitations; membrane prob-lems; low-volume prod.; cell growth problems (agita-tion/aeration?)	Innovative design improvements to achieve higher yields; scale-up growth control

flows downward to regions of great hydrostatic pressure, and there is a greater driving force for oxygen transfer. Air and spent gas expand in the rising leg and cause the fluid to circulate.

Fluidized and packed bed reactors usually use biocatalysts encapsulated within a polymeric carrier or fixed onto a solid or porous support by a variety of methods, for example, covalent bonding, adsorption, and cross-linking. Hollow fiber biochemical reactors belong to the category of mem-brane bioreactors where the biocatalysts are permanently entrapped in a confined region or compartment. Continuous flow biochemical research reactors are compared in Table 3.1.

An alternative process is solid-state fermentation, an old concept, but modern designs have revived interest. Culture grows on solid nutrients such as bran through which air can percolate. One method has shallow piles of material turned manually or automatically every few hours to aid oxygen transfer, to mix, and to combat compaction. Molds grow on moist solids,

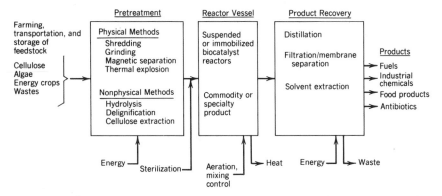

Figure 3.2. Flowchart of commercial bioconversion processes. (Kleinstreuer and Poweigha, 1984.)

and this may be more natural than submerged fermentation. High titers of enzymes have been produced at large scale from molds by solid-state fermentation.

The major cost factors for commercial bioprocesses are feedstock, bioconversion, and product recovery (Fig. 3.2). The production of intracellular proteins, antibiotics, vitamins, and certain acids are characterized by high recovery costs (Cooney, 1983), whereas the capital and operating costs for fermentation are of lesser importance. However, bioreactors usually play a central economic role in capital investment and a lesser role in operating cost.

The basic goal of production at minimum cost usually faces constraints such as limited biocatalytic capacity or upper bounds in momentum, mass, and heat transfer. Modern design objectives are

High product concentration

High yield and fast conversion of feedstock

Minimal by-product formation

Reactor flexibility for more than one process

Affordable aseptic techniques

Minimum capital investment and low maintenance

Effective control, that is, steady product quality and quantity

Easy separation of organisms from product stream

Minimal environmental impact

Reactors differ in features such as flow patterns and the mode of introducing the reactants. Three different schemes for bioreactor classification are in use. Atkinson (1974) considers the combination of the mode of substrate feed to the reactor and reactor geometry (batch, continuous, tubular packed

TABLE 3.2. Classification of Biochemical Reactors

	Enzyme Biocatalysts	Microbial Biocatalysts	
		Aerobic	Anaerobic
Homogeneous, liquid phase	BSTR[a] CSTR[a]	—	BSTR CSTR
Two-phase (Gas–liquid)	—	BSTR CSTR Bubble column Airlift (loop)	—
Immobilized biocatalysts	Hollow fiber Packed bed Fluidized bed	Airlift (loop) (Packed bed) (Fluidized bed)	Hollow fiber Packed bed Fluidized bed
Entrapped biocatalysts	BSTR CSTR Hollow fiber Membrane	—	BSTR CSTR Hollow fiber Membrane

[a]BSTR, Batch stirred-tank reactor; CSTR, continuous stirred-tank reactor. Both relative to liquid phase.

bed, and tubular fluidized bed). This classification is used in traditional chemical reactor engineering. A second system of classification is based on the configuration of the biocatalyst within the reactor (free, suspended, and immobilized biocatalyst). The third system is based on the mode of providing mixing within the reactor (Schügerl, 1982; Blenke, 1979).

A fourth system of classification, based on the type of biocatalyst used, namely, enzyme, aerobic microbe, or anaerobic microbe, is introduced here. This system automatically provides information about the phases that are present and when combined with the biocatalyst configuration leads to the scheme depicted in Table 3.2. This scheme of classification includes recent trends in the development of biorectors with whole cells or enzymes, immobilization techniques, and membrane technology. Attaching cells to a support in a reactor achieves very high cell density. A reactor packed with cells may experience mass transfer limitations for nutrients and for release of carbon dioxide. Heat transfer could also pose a problem in a commercial reactor since high growth or production rates could result in considerable evolution of heat. In two-phase aerobic reactors, mixing may be provided by the air alone through a sparger (as in bubble columns and airlift reactors) or together with a mechanical mixer [as in the batch stirred-tank reactor (BSTR) and continuous stirred-tank reactor (CSTR)]. Aerobic reactors with immobilized biocatalyst on the other hand, are three-phase, gas–liquid–solid, reactors in which mixing is provided solely by the gas.

Kleinstreuer and Poweigha (1984) have reviewed the modeling of the process dynamics of the major classes of biochemical reactors. Bailey (1980) also covers aspects of biochemical reaction engineering and biochemical reactors.

Immobilized enzymes or immobilized cells can function in conjunction with another bioreactor. One example is the hydrolysis of lignocellulosic materials with cellulase enzymes. The enzymes in the cellulase complex tend to have an affinity for cellulose. It is not very effective to immobilize enzymes that act on cellulose because reaction rates tend to be slow when bringing together a large substrate molecule and an enzyme tethered to a large support. One component of cellulase, called β-glucosidase, acts on a relativley small molecule, cellobiose, to split it into two glucose molecules. Immobilized β-glucosidase can speed the overall rate of cellulose hydrolysis by rapid removal of cellobiose that exerts feedback inhibition of cellulases.

In general, there are significant differences in the analysis and application of biocatalytic systems using enzymes on the one hand and microorganisms on the other. Enzymes resemble synthetic chemical catalysts. Applications are currently limited to cases involving a single enzyme, since the technology of multienzyme systems is either nonexistent or at best very rudimentary (e.g., Buchholz, 1982). Such one-enzyme applications include polysaccharide hydrolysis and deracemization of mixtures containing both the D-and L-isomers of optically active compounds. Other factors that limit the application of enzymes include their fragility with respect to shear forces and low tolerance for high temperatures. Microorganisms, on the other hand, are life forms that grow, and their catalytic activity must be monitored in concert with their growth processes. In addition, microbial biocatalysts usually contain the full sequence of enzymes required for complicated conversions. Enzymes and microorganisms survive and function only in a limited range of temperatures. Accurate and realistic modeling of microbial growth and/or enzyme kinetics are of great importance in the analysis and simulation of bioreactors. An overview of existing microbial population models is given in Figure 3.3.

In spite of the relative advantages and disadvantages of each type of biocatalyst, the goal in applications is to maximize the catalytic activity. This generally calls for high concentrations and may be achieved by immobilization, pelletization, and entrapment within a suitable region. In some applications, for example, waste treatment (Rittmann and McCarty, 1980), microbial films may form quite naturally. The analysis of biocatalytic systems may, thus, often require the simultaneous consideration of the inherent kinetics along with the mass transfer effects introduced by the biocatalyst configuration and the appropriate flow fields.

The literature on industrial reactors is not as extensive as that on laboratory devices, probably for proprietary reasons (Sittig, 1983). However, new designs may compete effectively with the traditional stirred-tank batch-operated reactor (e.g., Monsanto's hollow fiber perfusion reactor). It is clear that continuous industrial operation at high product concentration is essen-

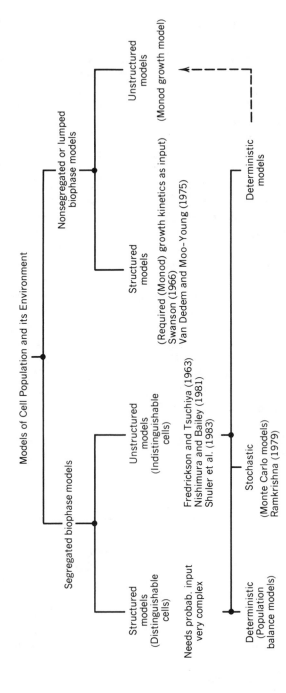

Figure 3.3. A classification of mathematical models of microbial populations. (Ramkrishna, 1979.)

tial for the new biotechnology. New reactor designs also aim at high capacity, simple technology, reliability, and low energy expenditure.

2. GENERAL MODELING CONCEPTS

The majority of bioreactor simulation models are oversimplified (Kleinstreuer and Poweigha, 1984). For example, rheological and nonisothermal effects are generally ignored and biokinetics and the temporal and spatial resolutions of the dependent variables (i.e., concentrations of substrate, nutrients, microorganisms, and products) are often poorly represented. This is understandable because reliable and complete data sets are lacking, and there are numerical solution constraints since the governing equations and necessary submodels are seldom available. Better representation is needed for the fluid mechanics (e.g., two- and three-phase flows, and rheology of fermentation broths), transient mass transfer mechanisms and biochemical reactions in stirred reactor trains with recycle, airlift fermenters, hollow fiber biochemical reactors, and fluidized as well as packed bed bioreactors.

Time pressures, availability of trained people, budgets, or insufficient information may restrict the analysis of a reactor system. Theoretical considerations for a new reactor design may be only a few hand calculations using simple objective functions and semiempirical relationships for a rough estimate of the rate-limiting step, volumetric flow rate, rate of feeding nutrients, residence time, mixing intensity, and gas requirements. A number of basic definitions, descriptions of the type of models available, and various modeling approaches are needed. Applications of these research tools are given throughout this chapter. Modeling concepts are discussed in conjunction with the definitions of various types of mathematical models, including supportive data analyses and special-purpose simulation languages. Different empirical and deterministic models are then discussed to illustrate the development and use of the basic modeling concepts.

2.1. Data Analysis and Parameter Estimators

Hampel (1979) gives an overview of the application of microcomputers interfacing with bench-top bioreactors for data reduction and analysis. The various steps in data analysis are discussed by Box and Jenkins (1976) and Vemuri (1978) among others. In *regression analysis,* given data sets are evaluated (usually with a statistical package) for the relative significance of all terms in a postulated equation. The most important terms are retained, and their coefficients are evaluated using the least-squares routine, for example. The resulting equation, a lumped-parameter, empirical model, possesses the best continuous fit of the discrete data sets. This empirical modeling approach is widely used in parameter estimation and input/output (I/O) model design for complex subsystems or large ecosystems.

Hence, the procedure of establishing *empirical models via curve fitting* follows three basic steps: (1) postulating a functional form for the relationship between input and output variable, for example, an nth-order polynomial, $y = \Sigma_{i=0}^{n} \alpha_i x^i$, which requires *at least* $(n + 1)$ data points; (2) estimating the number of significant terms (regression analysis), that is, the degree of the polynomial in this example; and (3) determining the coefficients, here the α's, which yield an optimal fit through the data points (curve fitting).

A recent development in biotechnology is the application of computers in controlling the course of bioconversion processes or monitoring biochemical experiments and estimating the reactor state from real-time measurements. In control applications the most common practice has been on–off control of physical variables of the biochemical process environment, for example, temperature, flow rates, and foam (Wang et al., 1979; Aiba et al., 1973). This has been feasible because quick-response sensors are generally available for monitoring these variables in real-time.

By comparison, computer control applications involving the use of feedback loops based on measurements of the chemical environment are less widespread, although such methods are currently under study (Bailey, 1981; Bowski et al., 1981; Stephanopoulos and San, 1981; Kiparissides and Mac-Greggor, 1981). The reason for this is the general unavailability of quick-response sensors of the chemical environment of biochemical processes. Among the chemical variables that are currently measurable in real time are oxygen and carbon dioxide concentrations in the influent and effluent gas streams of aerobic biochemical reactors, dissolved oxygen concentration, and pH (Wang et al., 1979). Stephanopoulos and San (1981) presented a mathematical analysis of the state estimation problem and its application to stirred-tank reactors. The underlying procedure in closed-loop control or on-line parameter estimation is (1) measurement in real-time of some variables through which the status of the biochemical environment can be determined, (2) postulation of some model of the system dynamics, and (3) implementation of an optimization process to minimize an appropriate objective function. In parameter or state estimation, a typical tool is the Kalman–Bucy filter (Bravard et al., 1981).

The status of computer control may be summarized as follows:

1. The availability of a larger number of quick-response sensors (in real time) for the chemical–biological environment would greatly expand applicability (Guilbault, 1981).
2. Current state and parameter estimation procedures have been applied successfully only to nonproduct-forming biochemical reactions and stirred-tank reactors (Stephanopoulos and San 1981); there is the need to extend the method to more general biosynthetic reactions and biochemical reactors.
3. Computer control is a downstream operation in the sense that it requires the existence of an actual biochemical reactor with the com-

puter and its associated data acquisition paraphernalia. It can only complement current off-line reactor modeling that is very much needed for design, scale-up studies, and the evaluation of process economics.

2.2. Special-Purpose Simulation Languages and Software Packages

Quite a number of useful *simulation languages* for common digital computers were developed by manufacturers of computer hardware, universities, and consulting firms: CSMP (IBM), CSSL-IV (Simulation Services), DYNAMO (MIT), GASP (Cellier, ETH, Zurich and Pritsker, Purdue University), GPSS-V (IBM), SIMSCRIPT-II.5 (CACI, Inc.), SLAM (Syst. Publ. Corp.) to name a few. Simulation packages facilitate routine modeling work especially in the simulation of first- and second-order rate equations (dynamic systems and classical process control), teaching dynamic simulation, reactor control, macroeconomics, statistical analyses, synthetic data generation, and systems optimization. In addition to these multipurpose computer languages, countless *software packages* are available that aid in model development and simulation (e.g., IMSL, 7500 Bellaire Blvd., Houston, TX). The subroutines are generally written in FORTRAN IV or PASCAL for better portability.

A unique and powerful tool is the *hybrid computer,* which is mainly used in mechanical systems dynamics, aircraft/space industry, chemical reaction kinetics, and process control. It consists of an analog computer that is interacting with a digital computer. The strength of the analog part lies in the *real-time* solution of multiple sets of nonlinear (ordinary) differential equations and the immediate, graphical display of the principal variables and their derivatives. The digital part, usually a minicomputer, handles I/O, control, data reduction, and most of all, program storage.

If the complexity of the modeling equations exceeds the capability of simulators or if more flexibility is desired, the differential–integral equations representing the process dynamics have to be solved with techniques discussed elsewhere (e.g., Roache 1972; Vemuri and Karplus 1981, Anderson et al. 1984).

3. SOME SIMPLE MODELING EQUATIONS AND THEIR APPLICATIONS

Before the basic conservation equations for bioreactor process simulation are introduced, a few simple modeling equations are discussed for basic bioreactor design, power requirement in a stirred tank, substrate uptake in a hollow fiber bioreactor, and cost estimates for catalytic reactors with separator.

3.1. Modeling Equations for Bioreactor Design

Cooney (1983), Wang et al. (1979), and others have presented a number of useful formulas for continuous flow processes. They define the objective function to maximize the ratio of product per unit of biocatalyst as an integral expression:

$$R_{p/x} = Y_{p/s} \int_0^{t_c} S_a(t)\, dt \tag{3.1}$$

where $Y_{p/s}$ is the conversion yield of raw material to final product that reflects selectivity, that is, the performance of the biocatalyst(s); $S_a(t)$ is the specific activity of the cells or enzymes that together with t_c, the half-life of the catalyst, have to be sustained at high values. Obviously, clever bioreactor design and operation help avoid catalyst inactivation, local substrate diffusion limitations, instabilities, and high shear stress for microorganisms so that high values for $R_{p/x}$ can be achieved. Typical numbers for $R_{p/x}$ range from 1 g of product per gram of catalyst for specialty products (e.g., drugs) to over 2500 for commodity products (e.g., corn syrup). Another handy formula for continuous, steady processes expresses the relationship between product concentration, P, and operating conditions such as the volumetric flow rate F:

$$P = \frac{S_a \Psi X}{F} \tag{3.2}$$

where Ψ is the reactor volume and X is the concentration of microorganisms.

Continuous flow reactors often have low product concentration when a high volumetric throughput is required. For aerobic fermentation under steady-state conditions, the active cell concentration depends on the rate-limiting oxygen transfer step:

$$X = \frac{Y_{O_2}}{\mu} k_L a(c^* - c) \tag{3.3}$$

Biological constraints are given by Y_{O_2}, the oxygen yield for growth and product formation and by μ, the specific growth rate coefficient. The rate of oxygen transfer is largely determined by k_L, the liquid film mass transfer coefficient, and by a, the total bubble surface area.

Finally, nutrient use and growth of microorganisms without inhibition are coupled via a nonlinear first-order rate equation that can be written for a transient process as follows

$$\frac{dS}{dt} = -\frac{1}{Y}\hat{\mu}(T)\frac{S}{K_s - S}X + \frac{F}{\Psi}(S_o - S) - \left(m + \frac{q_p}{Y_{p/s}}\right)X \tag{3.4}$$

where S is the concentration of organics, $\bar{Y}$ is an average growth yield, $\hat{\mu}$ is the maximum specific growth rate, K_s is the saturation constant, m is the maintenance requirement, and q_p is the specific rate of product formation.

3.2. Power Requirement in a Stirred Tank

Models of complex systems are often based on empirical data and arguments of dimensional analysis. Brauer (1979) distinguishes several functions for a stirrer in a reactor: energy transfer to the fluid, gas dispersion in a liquid, separation of gas and liquid, and mixing of all components of the process solution. The most important dimensionless numbers that govern power requirements in aerated stirred-tank reactors are the power or Newton number

$$\text{Ne} = \frac{N}{n^3 \, d_r^5 \rho} \tag{3.5}$$

and the Reynold's number

$$\text{Re} = \frac{n d_r^2 \rho}{\mu} \tag{3.6}$$

where N is the energy transferred by the mixer to the fluid, n is the number of revolutions of the stirrer per unit time, d_r is the diameter of the stirrer, ρ is the density, and μ is the viscosity of the fluid.

The Newton number is a function of Re, diameters, heights, lengths, thicknesses, number of blades, and the like.

Based on empiricisms and dimensional analysis, Brauer (1979) postulated

$$\text{Ne} = \frac{C_1}{\text{Re}} + \frac{1}{C_2 + C_3} - \frac{1}{C_2 + C_4 \text{Re}^2} + \frac{\text{Re}}{C_5 + C_6 \text{Re}^{1.113}} \tag{3.7}$$

where the C's are functions of the system parameters.

3.3. Substrate Uptake in a Hollow Fiber Biochemical Reactor

Consider steady convection down a porous tube with radial diffusion and suction and with first-order conversion of substrate c in a tubular reactor [e.g., a hollow fiber biochemical reactor (HFBR)] (Waterland et al., 1974).

Diffusion and reaction in the spongy matrix is decribed by

$$\frac{D_m}{r} \frac{\partial}{\partial r} \left(r \frac{\partial c_m}{\partial r} \right) = \frac{V_{\max}}{K_m} c_m \tag{3.8}$$

For slow kinetics, that is, small Thiele modulus that implies that $\lambda^2 < 1$, the flowing substrate solution in the lumen has a radially uniform concentration profile so that at the fiber wall

$$r = a \qquad \frac{v_0 a}{4} \frac{dc}{dz} = D_m \frac{\partial c_m}{\partial r}\Big|_a \text{ (equal flux across membrane)} \qquad (3.9a)$$

Additional boundary conditions are

$$r = r_0 \qquad \frac{\partial c}{\partial r} = 0 \text{ (no flux at outer fiber boundary)} \qquad (3.9b)$$

$$z = 0 \qquad c = c_0 \text{ (``initial'' condition)} \qquad (3.9c)$$

$$r = 0 \qquad \frac{\partial c}{\partial r} = 0 \text{ (symmetry along lumen axis)} \qquad (3.9d)$$

An analytical solution of the problem is

$$\frac{\text{exit concentration}}{\text{inlet concentration}} \qquad \frac{c}{c_0} = \exp(-4B\lambda\zeta) \qquad (3.10)$$

where the Thiele modulus $\lambda^2 = V_{max} a^2/K_m D_s$, the dimensionless axial distance

$$\zeta = \frac{zD_s}{a^2 v_0}$$

with

$$v_0 = \frac{2Q}{a^2 \pi N}$$

and

$$B = \frac{-K_1(\lambda\psi)I_1(\lambda) + I_1(\lambda\psi)K_1(\lambda)}{K_1(\lambda\psi)I_0(\lambda) + I_1(\lambda\psi)K_0(\lambda)}$$

with K and I being Bessel functions and $\psi = r_0/a$, where r_0 is the outside radius of the hollow fiber, a is the lumen or inside radius, and N is the number of fibers.

With $\psi = 1.75$ and $0 < \lambda < 0.5$, that is, kinetic control, B becomes $B \approx 1.031 \lambda$ so that

$$\ln \frac{c}{c_0} = -2.06 \frac{V_{max}}{K_m} \frac{za^2 \pi}{Q} \qquad (3.11)$$

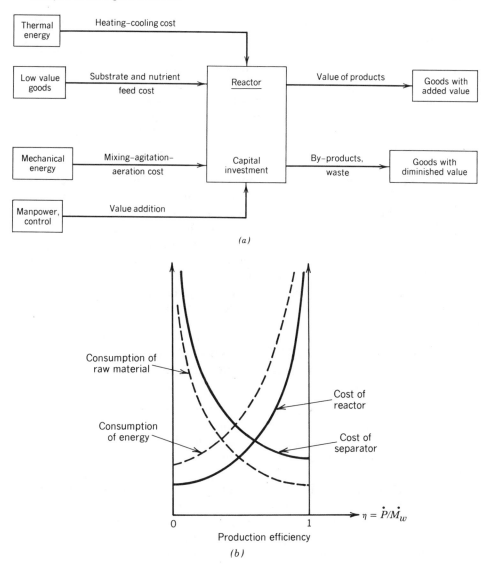

Figure 3.4. (a) Reactor value balance; (b) cost and consumption curves as a function of production efficiency. (Adapted from LeGoff, 1980.)

Hence, we can see that $c(z)$ reduces exponentially $\ln c/c_0 \sim \Psi_{\text{reactor}}$ and Q^{-1} (cf. Fig. 3.4).

3.4. Value Balance for Catalytic Reactor with Separator

Consider an industrial system that converts raw material with a low intrinsic value into more useful products. Figure 3.4a shows a *value balance* for a reactor (LeGoff, 1980). Material and energy streams are multiplied by specific dollar values. The system is a "black box" that carries out all necessary mechanical, chemical, and biological transformations. LeGoff (1980) discussed a catalytic reactor with associated separator for which the consumption and cost curves versus production efficiency are given in Figure 3.4b. Reactor yield is product concentration times flow rate. Formation step and purification are interrelated. Although formation may be optimized for cost at a low conversion efficiency, the larger amount of nonconverted reactant will result in an increase in the size of the separator and therefore in its cost. If the ratio $\dot{P}/\dot{M}_w$ (production versus waste stream) is large (i.e., very selective reactor), separation will be favored. But for many cases (e.g., if P and M_w are generated in competing reactions) an increase in selectivity can cost dearly (either in energy or capital equipment costs), and a new optimum should be sought. If the reactor is considered separately, its efficiency is a decreasing function of the mechanical energy dissipated in the reactor. But if the energy consumed in the separator and the recycle loop is taken into account, η is an increasing function of the total amount of degraded energy. The best compromise should be sought between a low efficiency using less energy but a lot of raw material (for a given value of P) and a greater efficiency that has the opposite effect.

4. GOVERNING EQUATIONS FOR SIMULATION OF REACTOR PROCESSES

A force, material, or energy balance over an infinitesimally small (reactor) volume, the representative elementary volume (REV), will lead to a set of coupled partial differential equations suitable for a distributed parameter approach. In contrast, a global balance over the entire reactor volume or its parts leads to a set of integral equations that can usually be reduced to (first-order) rate equations.

For the differential approach, the equations of motion and continuity, the species (or solute) mass transport equations, the energy equation, and the equation of state including the biokinetic and species flux expressions are of interest. Larger "particles" (e.g., biocatalyst beads in fluidized beds) that do not follow the fluid motion are usually represented by monodispersed spheres using Newton's second law of motion. Hence, the field equations in terms of fluxes for a generalized reactor system can be written as

(fluid continuity)

$$\frac{\partial \rho}{\partial t} + \nabla \cdot \rho \mathbf{v} = 0 \qquad (3.12)$$

(motion)

$$\frac{\partial}{\partial t}(\rho \mathbf{v}) + \nabla \cdot \rho \mathbf{v}\mathbf{v} = -\nabla p - \nabla \cdot \boldsymbol{\tau} + \sum \mathbf{f} \qquad (3.13)$$

(mass transport)

$$\frac{\partial c}{\partial t} + (\mathbf{v} \cdot \nabla)c = -\nabla \cdot \mathbf{j}_c \pm \sum S_c \qquad (3.14)$$

(internal energy)

$$\frac{\partial}{\partial t}(\rho c_p T) + (\mathbf{v} \cdot \nabla)\rho c_p T = -\nabla \cdot \mathbf{q}_H - \boldsymbol{\pi} : \nabla \mathbf{v} + \sum \bar{H}r \qquad (3.15)$$

The stress tensor $\boldsymbol{\tau}$, the net sinks or sources S_c such as biochemical reactions, cell growth, oxygen transfer, and so on, as well as the mass and heat flux ($\mathbf{j}_c$ and $\mathbf{q}_H$) have to be specified for the system in terms of the principal variables $\mathbf{v}$, c, and T in order to gain closure (see discussion below). An equation of state $\rho = \rho\,(p, c, T)$ is necessary when the assumption of incompressible flow does not hold. Suitable submodels for $\boldsymbol{\tau} = \eta \cdot \dot{\boldsymbol{\gamma}}$ have to be found when the suspension flow exhibits non-Newtonian rheological properties. The energy dissipation term, $\boldsymbol{\pi} : \nabla \mathbf{v}$ with the total stress tensor $\boldsymbol{\pi} = p\boldsymbol{\delta} + \boldsymbol{\tau}$, couples the energy equation with the momentum equation explicitly. The term $\Sigma \bar{H}r$ depicts thermal energy release by biochemical reaction, where $\bar{H}$ is the mean enthalpy and r is the reaction rate.

In practical applications, Eq. (3.15) is usually ignored and semiempirical expressions for single or two-phase flow velocity profiles are postulated in order to avoid solving Eqs. (3.12) and (3.13). Hence, the principal modeling equation for simulating the reactor process dynamics is Eq. (3.14) and its spin-offs. For example, transient convection–diffusion of species c_i in a multispecies two-phase flow field can be described by

(gas phase)

$$\frac{\partial(\phi c_i)}{\partial t} + (\mathbf{v}_G \cdot \nabla)c_i\phi = D_{G,i}\,\nabla^2(\phi c_i) + R_{G,i} + \sum_j I_{\mathrm{GL},j} \qquad (3.16)$$

(liquid phase)

$$\frac{\partial}{\partial t}[(1 - \phi)c_i] + (\mathbf{v}_L \cdot \nabla)(1 - \phi)c_i = D_{L,i}\nabla^2[(1 - \phi)c_i] + R_{L,i} - \sum_j I_{LG,j}$$
$$(3.17)$$

where ϕ is the fraction of gas volume per unit volume of the system, R is the reaction term reflecting the biokinetics, and I represents the interfacial mass transfer processes. Current research efforts concentrate heavily on realistic submodels for R and I (e.g., Poweigha, 1987; Prokop, 1983; Schügerl, 1982). The use of the modeling equations is discussed in the next sections on simulation case studies.

In contrast to the differential balance where the REV $\rightarrow 0$ and subsequently the field Eqs. (3.12) to (3.15) are produced, a *macroscopic balance* of forces, mass, energy, and/or entropy for a finite control volume (e.g.,

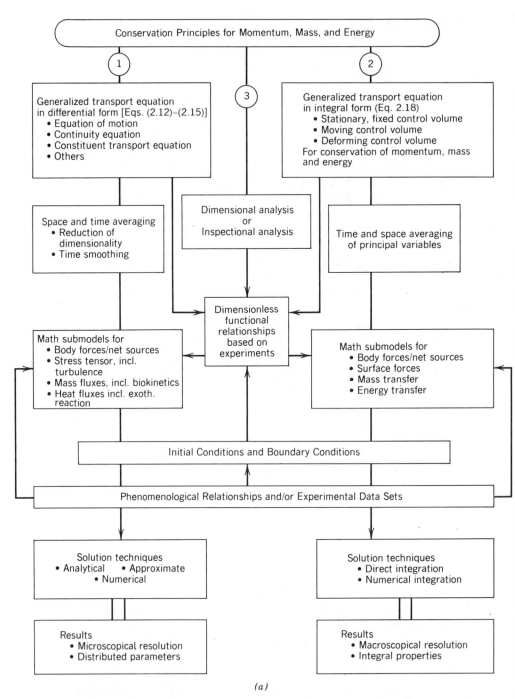

Figure 3.5. (a) Flowchart of general approach to deterministic modeling. (b) Model components and features.

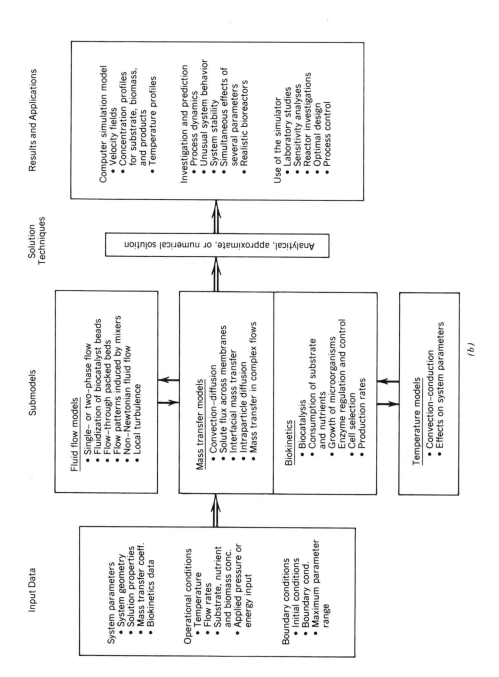

Input Data

System parameters
• System geometry
• Solution properties
• Mass transfer coeff.
• Biokinetics data

Operational conditions
• Temperature
• Flow rates
• Substrate, nutrient and biomass conc.
• Applied pressure or energy input

Boundary conditions
• Initial conditions
• Boundary cond.
• Maximum parameter range

Submodels

Fluid flow models
• Single– or two–phase flow
• Fluidization of biocatalyst beads
• Flow–through packed beds
• Flow patterns induced by mixers
• Non–Newtonian fluid flow
• Local turbulence

Mass transfer models
• Convection–diffusion
• Solute flux across membranes
• Interfacial mass transfer
• Intraparticle diffusion
• Mass transfer in complex flows

Biokinetics
• Biocatalysis
• Consumption of substrate and nutrients
• Growth of microorganisms
• Enzyme regulation and control
• Cell selection
• Production rates

Temperature models
• Convection–conduction.
• Effects on system parameters

Solution Techniques

Analytical, approximate, or numerical solution

Results and Applications

Computer simulation model
• Velocity fields
• Concentration profiles for substrate, biomass, and products
• Temperature profiles

Investigation and prediction
• Process dynamics
• Unusual system behavior
• System stability
• Simultaneous effects of several parameters
• Realistic bioreactors

Use of the simulator
• Laboratory studies
• Sensitivity analyses
• Reactor investigations
• Optimal design
• Process control

(b)

51

reactor) generates an integral equation (Reynolds transport theorem):

$$\frac{D}{Dt} G_{\text{syst}} = \frac{\partial}{\partial t} \iiint_{\text{cv}} \eta\rho d\Psi + \iint_{\text{cs}} \eta\rho\mathbf{v} \cdot d\mathbf{S} \qquad (3.18)$$

$$\begin{Bmatrix} \text{Time rate of} \\ \text{change of arbitrary} \\ \text{property G as} \\ \text{the system moves} \end{Bmatrix} = \begin{Bmatrix} \text{Accumulation of} \\ \text{specific property} \\ \eta \text{ within control} \\ \text{volume} \end{Bmatrix} + \begin{Bmatrix} \text{Net efflux of} \\ \text{specific property} \\ \eta \text{ through} \\ \text{control surface} \end{Bmatrix}$$

where $G_{\text{syst}} = \iiint_{\Psi\text{syst}} \rho\eta d\Psi$ and η is equal to G per unit mass.

The flowchart in Figure 3.5a (page 50), depicts three avenues to model a deterministic system, that is, pathway (1) relies on partial differential equations, pathway (2) is based on macroscopical balances, and pathway (3) uses dimensional analysis to develop functional relationships between a dependent engineering variable and important system parameters. A practical application of Figure 3.5a to the modeling of bioreactor process dynamics is depicted in Figure 3.5b (page 51), which is self-explanatory. Both parts of Figure 3.5 indicate the need for submodels and boundary conditions in order to gain closure. For example, Harder and Roels (1982) summarized microbiological and biochemical principles to be used as math submodels for mass fluxes as (1) direct mass–action law from which Monod's equation can be deduced, (2) regulation of the activity of enzymes that describes microbial control mechanisms, (3) regulation of the macromolecular composition of the cell, (4) natural selection within a population of a species, and (5) changes in the composition of a mixed species population. Shuler and Domach (1983) and Bailey (1983) presented coupled sets of first-order rate equations modeling the changing states of a typical single cell. Such structured-segregated models are useful for predicting changing biochemical mechanisms due to external disturbances.

5. COMPUTER SIMULATION STUDIES

In previous sections, simple mathematical models were presented for which analytical or approximate solutions are readily available. The governing equations describing the transport and conversion mechanisms of bioreactors were introduced in the last section; here two representative computer simulation studies are discussed.

5.1. Gas–Liquid Bioreactors (Case Study)

Airlift, tower or loop fermenters, and three-phase fluidized bed bioreactors have the advantages of less expensive oxygen transfer, no mechanical agita-tion, and easier scale up for certain configurations. The key parameters for gas–liquid reactor design and modeling are the mass transfer coefficient, gas holdup, axial dispersion coefficients, fluid fluxes, and the pressure requirement for sparging air into the reactor. The volumetric mass transfer coefficient, $k_L a$, and the rate of oxygen consumption versus power input are useful indexes of reactor performance.

Schügerl (1982) reviewed the oxygen transfer rate and cell mass productivity of yeast and bacteria in bench-scale tower loop reactors with regard to medium properties, construction parameters, and process variables. Hatch and Wang (1975) discussed experimental and theoretical aspects of oxygen transfer in airlift fermenters. Merchuk et al. (1980) presented a computer simulation model for the axial substrate, biomass, and oxygen concentration profiles in an airlift bioreactor.

Sparging filtered air into an airlift bioreactor sets up two-phase flow patterns where substrate, nutrients, microorganisms, and products are well agitated, and high oxygen transfer rates are achieved. Considerable backmixing in both phases, high pressure drop, bubble coalescence, and reduced oxygen transfer into highly viscous fermentation broths can be disadvantageous in some instances (Shah et al., 1982). The modeling steps of system conceptualization, development of problem-oriented equations and closure, parameter determination, computer solution, and discussion of results are illustrated using the work by Merchuk et al. (1980) as a basis.

For the airlift fermenter with two concentric cylinders and a centered sparger (Fig. 3.6), two-phase flow is assumed only in the riser. The head space is regarded as a CSTR. Biological activities occur in all three regions of the reactor. The fluid dynamics are simplified by postulating plug flow for both fluids (i.e., air and water) in the riser as well as in the downcomer, the annular liquid flow region.

The governing equations for the *riser and downcomer* are

(gas phase)
$$\frac{\partial(\phi c_i^G)}{\partial t} + \bar{u}_G \frac{\partial(\phi c_i^G)}{\partial z} = D_{G,i} \frac{\partial^2(\phi c_i^G)}{\partial z^2} + r_{G,i} - \sum_k I_{\mathrm{GL},k} \quad (3.19)$$

(liquid phase)
$$\frac{\partial[(1-\phi)c_i^L]}{\partial t} + \bar{u}_L \frac{\partial[(1-\phi)c_i^L]}{\partial z} = D_{L,i} \frac{\partial^2[(1-\phi)c_i^L]}{\partial z^2} + r_{L,i} - \sum_k I_{\mathrm{LG},k}$$
$$(3.20)$$

(energy equation)
$$\frac{\partial T}{\partial t} + (\mathbf{v} \cdot \nabla)T = \alpha \nabla^2 T + \sum_i r_{L,i}(\Delta H)_{L,i} + \sum_k r_{L,k}(\Delta H)_{L,k} \quad (3.21)$$

where c is the concentration of substrate, biomass, or oxygen; ϕ is the gas holdup; $\bar{u}$ is the averaged axial velocity; and T is the fluid temperature. The

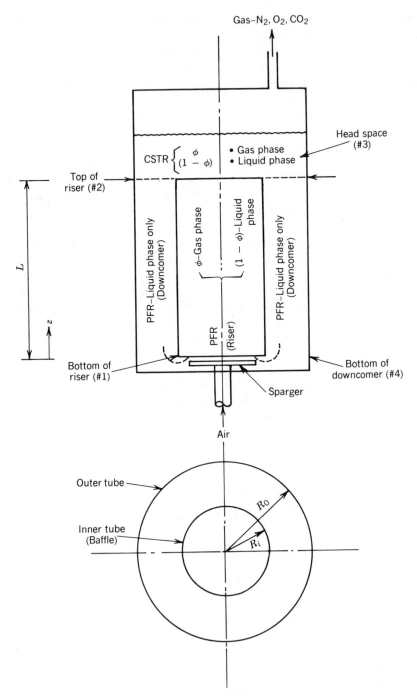

Figure 3.6. Schematics and conceptualization of airlift fermenter. (Kleinstreuer and Poweigha, 1984.)

terms r and I represent homogeneous reactions and interfacial mass transfer, respectively. These equations describe the multispecies mass and energy transfer in two-phase plug flow reactors with axial dispersion.

For the *head space*, which is regarded as a continuously stirred-tank reactor, the conservation principles can be stated as follows:

(gas phase)

$$\frac{d}{dt}(V_{SG}\phi c_i^G) = F_{G,\text{in}}(c_{i,\text{in}}^G \phi) - F_{G,\text{out}}(c_i^G \phi) + V_{SG}r_{G,I} + \sum_k I_{LG,k} \quad (3.22)$$

(liquid phase)

$$\frac{d}{dt}[V_{SL}(1 - \phi)c_i^L] = F_{L,\text{in}}[c_{i,\text{in}}^L(1 - \phi)] - F_{L,\text{out}}[c_i^L(1 - \phi)]$$

$$+ V_{SL}r_{Li} - \sum_k I_{LG,k} \quad (3.23)$$

(energy)

$$\frac{d}{dt}[(V_{SG} + V_{SL})T] = F_{T,\text{in}}T_{\text{in}} - F_{T,\text{out}}T + V_{SG}\left[\sum_i r_{Gi}(\Delta H)_{Gi}\right]$$

$$+ V_{SL}\left[\sum_k r_{L,k}(\Delta H)_{L,k}\right] \quad (3.24)$$

Equations (3.19) to (3.24) are equivalent to those presented by Cichy et al. (1969) for two-phase tubular reactors and by Schaftlein and Russell (1968) for two-phase stirred-tank reactors.

The simplifying assumptions incorporated by Merchuk et al. (1980) are steady-state operation, constant gas holdup throughout the riser and head space, no axial dispersion, isothermal processes, and axial dependence only of principal variables.

The assumption of constant gas holdup, ϕ, deserves comment. A few researchers have reported both radial and axial variations of ϕ in bubble columns (Shah et al., 1982), which are similar to airlift reactors. Merchuk and Stein (1981) have proposed a procedure for measuring the axial variation of ϕ in the airlift reactor.

For aerobic reactions, the assumption of isothermal biokinetics would be unrealistic for such exothermal reaction except that fermentations generally have good temperature control. Appropriate reaction and interfacial mass transfer terms are needed to gain closure. System variables considered are concentrations of substrate, biomass, and oxygen in the liquid phase and concentration of oxygen in the gas phase.

Reaction stoichiometry gives

$$\frac{1}{Y_s}[S] + \frac{1}{Y_{\text{ox}}}[O_2] = \text{cell mass } (B) + \text{products } (P) \quad (3.25)$$

$$r_B = \frac{\mu[B][S]}{K_m + [S]} \quad \text{(Monod kinetics)} \tag{3.26}$$

$$r_s = -\frac{1}{Y_s} r_B \tag{3.27}$$

$$r_{O_2} = \frac{Y_s}{Y_{ox}M_{ox}} r_s \tag{3.28}$$

Equation (3.28) corrects an equation of Merchuk et al. (1980). Although product formation kinetics have not been considered, the primary metabolite concentration profile should be proportional to the substrate concentration profile. Product formation in the case of secondary metabolites has to be considered separately.

The oxygen transfer rate (Danckwerts, 1970)

$$I_{\text{LG,ox}} = -k_L a \frac{P_{O_2}}{H} - c \tag{3.29}$$

results from the standard two-film interfacial mass transfer model. The overall oxygen transfer rate (Kuraishi et al., 1975) is

$$k_L a = 47.8 \, J_G^{0.997} \tag{3.30}$$

System hydrodynamic considerations give (Hatch, 1973)

(gas volume flux)
$$J_G = \phi \cdot V_G = \frac{Q_{\text{gas}}}{A_{\text{riser}}} \tag{3.31}$$

(liquid volume flux)
$$J_L = (1 - \phi)V_L \tag{3.32}$$

(superficial gas velocity)
$$V_G = 1.065 J + 32 \tag{3.33}$$

(superficial liquid velocity)
$$V_L = 0.0152 \, J_G 1.73 \tag{3.34}$$

The detailed equations are derived by Poweigha (1987).

The corresponding boundary conditions are

$$\text{at } z = 0 \quad \begin{aligned} c &= c_s \\ S &= S_s \\ B &= B_s \end{aligned} \tag{3.35}$$

For the downcomer, $z = 0$ corresponds to the head space and $z = L$ to the bottom of the downcomer.

The modeling equations are further simplified and written in a more compact form by the following:

1. Calculation of the fluid dynamics parameters ϕ, J_L, V_L, and V_G as a function of J_G by solving Eqs. (3.31) to (3.34) simultaneously
2. Direct integration of the cell-mass–substrate-balance equations for riser and downcomer
3. Nondimensionalization of the modeling equations using the results of (2) and defining

$$S_A = \frac{S}{S_f} \; ; B_A = \frac{B}{Y_s \cdot S_f} \; ; Y_1 = \frac{P_{O_2}}{P_{O_{2,0}}} \; ; Y_2 = \frac{c_{O_2}}{c_{O_2}} \; ; \eta = \frac{z}{L} \quad (3.36)$$

and the following dimensionless groups

stoichiometric group

$$D_1 = \frac{Y_s \cdot S_f H}{Y_{ox} M_{ox} P_{O_{2,0}}} \quad (3.37)$$

reaction group

$$R = \mu \frac{L}{J_L} \quad (3.38)$$

Stanton number, gas phase

$$St_G = \frac{L \cdot k_L a RT}{J_G H} \quad (3.39)$$

Stanton number, liquid phase

$$St_L = \frac{L \cdot k_L a}{J_L} \quad (3.40)$$

Inserting the dimensionless variables and system parameters yields the final modeling equations.

Merchuk et al. (1980) did not state the values used for Y_s, β, and P_{tot}, but reference was made to a study by Goldberg et al. (1976) for "the stoichiometric constants." Furthermore, Y_{ox} did not appear in their derivations. Our model uses values for Y_{ox} from Wang et al. (1979, p. 179). For β and P_{tot} we used 0.1 and 1.5 (atm), respectively.

Merchuk et al. (1980) used the reaction group $R = \mu L/J_L$ for sensitivity analyses. Clearly R depends on the independent variables J_G and L for a given microbial species, that is, a fixed μ. Different combinations of L and J_G may give the same value of R, but the J_G values may correspond to different flow regimes within the reactor, whereas L may not correspond to consistent reactor sizes. The combinations of the defining variables for R should be consistent. Steady-state and lumped parameter derivations were used, but they should not apply to other than steady-state system models. Non-

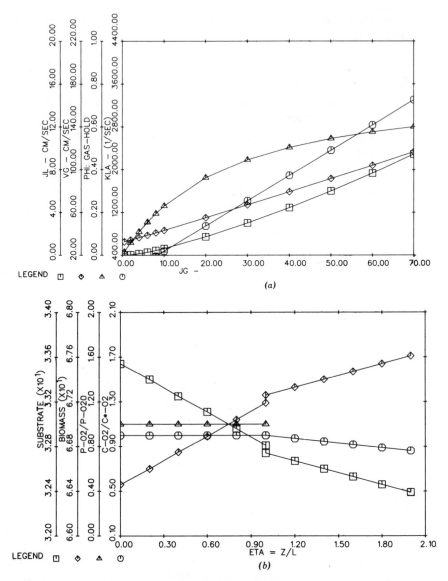

Figure 3.7. (*a*) Fluid dynamic and mass transfer parameters for conventional airlift fermenter (Merchuk et al., 1980); (*b*) axial variation of system state variables in airlift fermenter.

steady-state situations are the rule rather than the exception in most practical cases. This immediately exposes one of the major limitations of the model.

The results of the model implementation are summarized in Figures 3.7*a,b* and 3.8. Key input data for these runs are presented by Kleinstreuer and Poweigha (1984). Figure 3.8*a* presents the variation of key average fluid

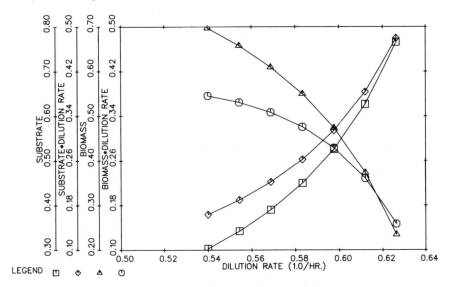

Figure 3.8. Substrate and biomass at bottom of downcomer.

dynamic–transport parameters within the reactor as a function of the gas volume flux, J_G. It is seen that J_L, V_G, and k_La all change almost directly with J_G; the gas holdup, ϕ, on the other hand, appears to taper off with J_G. As we discussed earlier, the hydrodynamic submodel used is quite idealized; one of the reasons is the difficulty of adequately describing the two-phase flow hydrodynamics within the reactor. It is expected that further increases in J_G would cause a discontinuity in the value of each parameter as the flow field changes into a different regime, for example, bubble flow into slug flow.

Figure 3.8b shows the axial variation of the nondimensionalized system state variables, substrate, S_A, biomass, B_A, gas phase oxygen partial pressure, Y_1, and dissolved oxygen, Y_2. The region 0–1.0 for η corresponds to the riser section of the reactor; the region 1.0–2.0 corresponds to the downcomer. Ranges selected for each variable are consistent with physical expectations. For example, the substrate level decreases progressively from the bottom of the riser to the bottom of the downcomer because of consumption in microbial growth. The discontinuity at η = 1.0 in each of the variables represents the conceptualization of the head space region as a CSTR; we have, in effect, a plug flow reactor, PFR-CSTR-PFR train corresponding to the riser-head space-downcomer sequence. Both Y_1, and Y_2 change only slightly and are virtually constant in the riser section. This results from a low metabolic rate of the microorganisms with a correspondingly low oxygen demand. It is assumed that all free oxygen is expelled from the head space region; thus only the Y_2 profile is indicated in the downcomer region. In this region, the higher biomass concentration offsets the reduced substrate concentration to cause a higher demand for oxygen, as indicated by the perceptible change in Y_2. A different set of basic input parameters would, of course, change the actual variations in each of the state variables, although the qualitative trends would remain the same.

Figure 3.9, which corresponds to Figure 5 of Merchuk et al. (1980), shows the variation of the substrate, S_A, and biomass, B_A, at the bottom of the downcomer with dilution rate D. Also indicated are the products S_A*D and B_A*D, which may represent the substrate consumption and biomass productivity, respectively. The trends in S_A and B_A are as expected. As D increases, there should be a correspondingly lower conversion of the substrate and lower biomass productivity. This is represented by high levels of unused substrate and low levels of effluent biomass, respectively.

Each parameter must lie within an appropriate range of values if a solution of the model equations is to be guaranteed. Typical indicators of inconsistent parameter values include the occurrence of negative variables for logarithmic function evaluations and the failure of the nonlinear equation solutions to converge. Examples encountered include (1) a maximum value of 775 mg/liter) for the substrate feed concentration, (2) a maximum value of 4 for ξ, the ratio of cross sections of riser to downcomer, and (3) a maximum value of 1000 cm for the height of the riser. These maxima correspond to particular choices of the other variables.

The problem of determining the boundaries of the acceptable parameter space, a priori, if possible, is challenging; however, in models in which the parameters are highly coupled with the system state variables, as we have here, a systematic search technique would be necessary to keep computer cost reasonable.

Avenues for the extension of this model are as follows:

1. The development of a non-steady-state submodel for microbial kinetics covering the entire range of the microbial growth curve.
2. The consideration of the interfacial mass transfer processes from the viewpoint of irreversible thermodynamics appropriate to the non-steady-state situation. The objective would be to obtain a submodel for $k_L a$, which incorporates system transients.
3. A hydrodynamic submodel for the two-phase flow that can describe both radial and axial variations in the flow parameters, for example, gas holdup, ϕ; in addition, rheological properties on non-Newtonian solutions should be incorporated.
4. Nonisothermal operation. The goal would be to account for the temperature dependence of the microbial kinetics and possibly of the mass transfer parameters. Since aerobic processes are in general exothermic, it is reasonable to expect that significant temperature changes would arise to negate the assumption of isothermal operation.
5. Inclusion of carbon dioxide transport. This would be useful since significant levels of the production of this gas accompanies the microbial growth, and the gas bears an important relationship to the ionic

state of the liquid environment. Such an inclusion would enable the calculation of pH effects on the microbial kinetics.

6. The use of reactor submodels that are intermediate, i.e., below the extreme cases of complete mixing and plug flow. The material is essentially available and only requires incorporation into the overall model of the airlift reactor.

5.2. Tower and Loop Bioreactors

Other types of gas–liquid bioreactors may be classified as tower and loop bioreactors. In these reactors, the flow field is essentially the same throughout the entire volume, whereas in an airlift reactor the flow field varies with location. Tower and loop bioreactors can be operated in batch mode with each stage regarded as a CSTR, that is, perfect liquid mixing prevails and no back flow or back mixing occurs.

Recent reviews on the operation of this class of bioreactors have included those by Schügerl et al. (1977) on bubble columns, Blenke (1979) on loop reactors, and Schügerl (1982) on the use of tower bioreactors for cell mass production. Shah et al. (1982) also presented a general review on the estimation of parameters for bubble columns; some of that information is directly relevant to bioreactor applications.

The key variables for modeling the two-phase flow dynamics and mass transfer of these reactors include the relative gas holdup, bubble sizes, superficial and mean liquid–gas velocities, volumetric flux densities, overall oxygen transfer coefficient, and power spectrum of air inlet system. The interactions of these variables together with the kinetics of a given biological system are, in general, described by a highly coupled, transient, nonlinear set of partial differential equations. However, because of computational constraints, current estimation methods use semiempirical correlations obtained for specific systems (Shah et al., 1982; Schügerl, 1982; Moo-Young and Blanch, 1981). For comparing the performance of different air-sparged systems, it is proposed to use in addition to the same biological system some form of power ratio, for example, oxygen consumption rate per unit compression horsepower (Hatch, 1975). Other recent accounts of the modeling of tower and loop reactors are by Luttmann et al. (1982a,b), Ziegler et al. (1977), and Russell et al. (1974). The modeling equations are similar to those for the riser section of the airlift reactor and will not be presented again. Applications of this class of bioreactors have been carried out for single-cell protein production from hydrocarbon substrates and biological treatment of wastewater (Schügerl, 1980).

5.3. Stirred-Tank Reactors

Biochemical reactor tanks are operated in three general modes: stirred-tank batch reactors, intermittently fed batch reactors, and continuous flow stirred-tank reactors. Several different configurations depend on the type and placement of biocatalysts and the number of phases present. A large body of knowledge exists concerning the operation of stirred-tank reactors. Several aspects of the theory have been well studied by Nauman (1969, 1981), Patterson (1981), Brodkey (1981), Leng (1981), and Hamer (1982) among others. Schaftlein and Russell (1968) presented a definitive study of two-phase tank-type reactors. For reviews of the applications of stirred tanks in biochemical engineering, see Ollis (1977), Bailey (1980), and Hamer (1982).

5.4. Completely Mixed Reactor Trains

Little published work is devoted exclusively to ideal reactor trains as used in biochemical processing. In most contexts the concept is used in characterizing the nonideal mixing of a single reactor by considering a series of N equal-sized ideal CSTRs. One of the cases in which the ideal CSTR-train model is used is the two-reactor sequential process for anaerobic digestion (e.g., Pohland and Ghosh, 1971) in which the volatile acids are generated in the first reactor by the acid-forming bacteria, whereas the methane-forming microorganisms use the effluent as the feed for the second reactor in completing the fermentation process. The sequential arrangement permits optimal operation of each step by independently controlling the detention time. This concept can be used in the general case for processing nongrowth-associated product-forming systems and in accommodating intermediate separation–biocatalyst recovery processes before final product recovery is implemented. A reasonably general STR train is presented in Figure 3.9. For nongrowth-associated product formation, the feed variables for reactor II

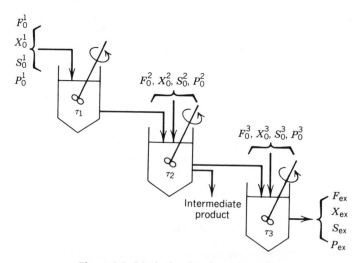

Figure 3.9. Ideal stirred-tank reactor train.

might be nonzero. For growth-associated product-forming systems, a single reactor would be adequate.

To model such a system, individual sets of conservation balances (for cells, substrates, and products) are written for each reactor, and the equations are solved as one system. For the system indicated, the governing equations are derived from the general, transient, variable-volume formulation for one STR (Dunn and Mor, 1975), which in turn can be deduced from a reduced form of Equation (3.18) given earlier.

Cells, X

$$\dot{X} = \frac{1}{V} (F_oX_o - F_1X) + r_x - \frac{(F_0 - F_1)}{V} X \qquad (3.41)$$

Substrate, S

$$\dot{S} = \frac{1}{V} (F_oS_o - F_1S) - \frac{r_x}{Y_{x,s}} - \frac{(F_0 - F_1)}{V} S \qquad (3.42)$$

Product, P

$$\dot{P} = \frac{1}{V} (F_oS_o - F_1P) + \frac{r_s}{Y_{x,p}} - \frac{(F_o - F_1)}{V} S \qquad (3.43)$$

Volume, V

$$\dot{V} = F_o - F_1 \qquad (3.44)$$

Note that in the case of equal rates of inflow and outflow, the last term of each equation drops out and the system simplifies to the case of a CSTR. Once written for each STR in the train, the system of dynamic equations may be solved to obtain the time-dependent profiles of the variables of interest, for example, microbial cells, substrates, and products in each reactor. In addition, the system parameters may be subjected to sensitivity analyses to obtain an optimal choice of the combination of residence times.

A model of a train of ideal stirred-tank reactors consisting of two CSTRs was developed and calibrated against experimental data for the two-phase anaerobic digestion of glucose (Cohen et al., 1979). Kleinstreuer and Poweigha (1982) developed a dynamic, one-reactor simulator for anaerobic digestion. The two-reactor train model used in this case study is an adaptation of the work of Kleinstreuer and Poweigha (1982).

The results of the two-reactor model implementation for the data of Cohen et al. (1979) are presented in Figures 3.10 and 3.11. The substrate feed profile was reported by Cohen et al. (1979). The key calibrating variables were the rates of CO_2 and H_2 production and their ratio for the acidogenic reactor and the yields of CH_4 and CO_2 for the methanogenic reactor. Gas-flow rates were not reported for the methanogenic reactor.

There is a reasonable match between the model and experimental data (cf., Cohen, et al. 1979) for the acidogenic reactor (rate of H_2 production and rate of H_2 production to rate of CO_2 production). The gas phase H_2 fraction

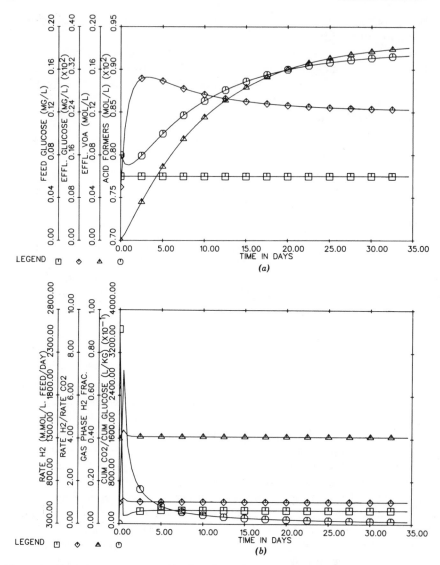

Figure 3.10. (*a*) Two-stage anaerobic digestion acidogenic phase–RXTOR #1(1/2); (*b*) two-stage anaerobic digestion acidogenic phase–RXTOR #1(2/2).

also matches the experimental result (about 50%). There is also a qualitative match between the model and the experimental results (gas phase CH_4 fraction and total gas produced per liter of influent). It should be noted, however, that there is a discernible time-lag effect between the states in the two reactors. Start-up transients died out in about 15 days in the acidogenic reactor; the corresponding time in the methanogenic reactor was about twice as long. Cohen et al. (1979) reported this observation. In general, the time

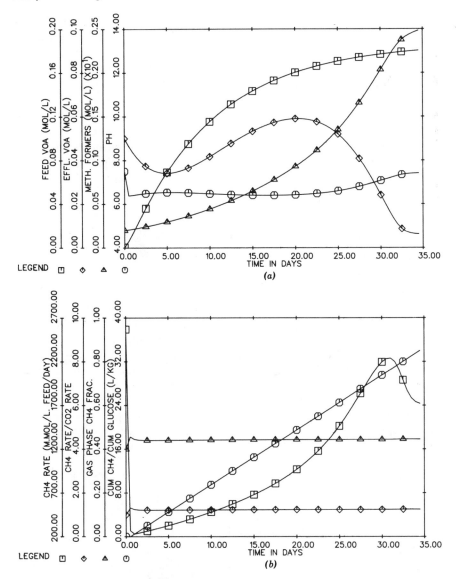

Figure 3.11. (*a*) Two-stage anaerobic digestion methanogenic phase–RXTOR #2(1/2); (*b*) two-stage anaerobic digestion methanogenic phase–RXTOR #2(2/2).

lag would depend on the initial conditions in the methanogenic reactor, its feed profile, and, of course, the kinetics.

A major difference between the Cohen et al. (1979) study and most other anaerobic digestion studies was the short retention time of 10 hr for the acidogenic reactor. This corresponds to a maximum specific growth rate of 2.4 day^{-1}; the values most often reported in the literature are much less (usually less than 0.5 day^{-1}). The model implementations thus used a 10-day retention time for the acidogenic reactor and a 20-day time for the methanogenic reactor.

A comparison between the two-reactor train and the single reactor for methane generation indicates that the gas production for the one-reactor process was higher than that for the two-reactor train. This arose from the high level of volatile acids production from the acidogenic reactor that was the feed for the methane-forming bacteria in the second reactor. The two-reactor train model was not in this case operated at the optimum hydraulic loading; the appropriate level should produce volatile organic acids of concentrations lower than those required to inhibit the methane-forming bacteria substantially in the methanogenic reactor.

5.5. Stirred-Tank Reactors with Recycle

A typical reactor train with recycle is depicted in Figure 3.12. However, the level of sophistication is low when nonideal mixing and recycle with time delay are not considered. The last feature is important in actual recycle situations when recovery of process water, energy, or cells plays a significant role. Recycle of the liquid, gaseous, and/or solid phase helps to improve yield, productivity, and bioconversion for most fermentation processes. Furthermore, recycle will have a place in fully optimized biomass conversion processes designed to satisfy economic and environmental restraints.

Lam and Bungay (1983) have a program for interactive computer graphic analysis of multistage continuous fermentation with provision for recycle or supplemental feeding at any stage. Monod coefficients are specified for each

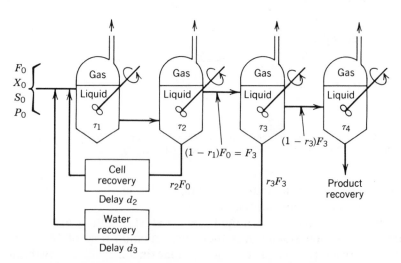

Figure 3.12. Typical ideal stirred-tank reactor train with recycle.

stage, and flow rates and reactor volumes can be varied. The computer displays cell concentration, substrate concentration, and productivity for each stage and for the overall system. A light pen selects menu options for rapid testing of many permutations and configurations. This program assumes perfect mixing and knowledge of the Monod coefficients. It would be a major advance to derive functional relationships in terms of substrate concentration and residence time so that the program could adjust the Monod coefficients itself.

5.6. Immobilized Biocatalyst Reactors

Immobilized bioreactors have a heterogeneous solid phase in which the biocatalyst is attached to a solid support or encapsulated in beads. The biocatalyst may be immobilized as free enzymes or whole cells that contain the active enzymes. The substrate, usually in an aqueous medium, is passed over the anchored catalyst. Reaction rates can be several times higher than in continuous flow stirred-tank reactors (CFSTR). The potential for significant increases in throughput is one major attraction of immobilized biocatalyst reactors, since this would translate into significant reductions in reactor volume. Immobilized biocatalyst systems have been used with all traditional reactor configurations: BSTR, CFSTR, fixed bed, fluidized bed, and tubular.

Traditional chemical reactor engineering provides a vast repertoire of analytical techniques for modeling immobilized biocatalyst reactors (Bailey, 1980). However, the adaptation is not usually straightforward because biocatalysts may superimpose several extraneous phenomena (relative to diffusional mass transfer) on the inherent kinetics. Various physical processes may have to be considered, and there are heat transfer problems more difficult than those that traditional chemical reactor engineering has essentially solved.

There are four major methods for immobilizing biocatalysts, namely,

1. Cross-linking within or on a suitable carrier
2. Covalent bonding to a polymeric matrix
3. Adsorption on or in a carrier
4. Encapsulation or entrapment

Both enzymes in vitro and whole cells (both viable and lysed) have been successfully immobilized (Messing, 1975). Method (4) applies especially to fibers and membranes, and with the hollow fiber reactor being a major example. The supports used for immobilization may be nonporous or porous. Members of the group of porous carriers include (1) controlled pore materials that are characterized by a narrow pore-size distribution, for example, controlled pore glass (CPG), (2) broad-pore distribution carriers, for example, alumina, and (3) gel structures.

Some of the extraneous phenomena are as follows (Trevan, 1980):

1. Conformational or steric effects that upset the normal three-dimensional configuration of the catalyst required for its activity (in the case of enzymes). This could lead to outright denaturation or deactivation of the catalyst.
2. Partition effects that arise from the use of electrically charged supports and/or substrates upset the pH dependence of the catalyst (enzyme) and introduce an electric field component to the normal diffusional mass transfer mechanism. In order to accommodate this higher level of complexity, appropriate electrostatic theory has to be indicated in the analysis.

The relative importance of these complications will depend on both the mode of immobilization and the type of carrier material, so immobilized biocatalyst reactors can be difficult to model.

Assumptions often used to simplify the analysis include

1. Plug flow or Poiseuille-type velocity profiles
2. Steady state
3. First-order irreversible reaction or Michaelis–Menten or substrate-inhibition kinetics
4. Isothermal operation
5. No time delay in case of recycle

Governing equations and boundary conditions depend on the type of immobilization and the structure of the carrier. The literature treats

1. Nonporous carrier and physically adsorbed catalyst: In this case, the biochemical reaction is coupled with external film diffusion. For charged carrier and/or substrate, electrostatic effects may be incorporated.
2. Porous carrier and any type of catalyst immobilization.
   a. Reaction is coupled with external and possibly intraparticle mass transfer
   b. Case (a) with electrostatic effects
   c. Case (a) with catalyst deactivation
   d. Case (c) plus electrostatic effects, that is, reaction coupled with external and introparticle mass transfer, catalyst deactivation, and electrostatic effects

5.7. Hollow Fiber Biochemical Reactor (Case Study)

A hollow fiber biochemical reactor has the biocatalysts (enzymes or cells) separated from the reactant by a semipermeable membrane (Fig. 3.13) that

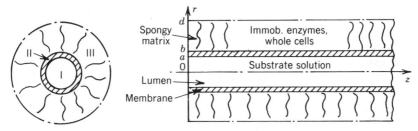

Figure 3.13. Schematics of single hollow fiber.

permits passage of substrate and products but restricts microorganisms. Entrapment of the catalysts, usually in the annular region, can be accomplished by using a carrier liquid in a spongy matrix, just a carrier liquid possibly as a countercurrent to the feed stream in the lumen, and a porous matrix for enzyme immobilization. Several geometries are used: planar, cylindrical (commonly used), and spherical (microencapsulation).

The advantages of HFBRs are little steric hindrance and deactivation for enzymes, high concentration of microorganisms, high volumetric productivity, basically no washout, and possibility of using multiculture conversion. All operations have problems with fouling, leakage, and diffusion limitations for substrates and products. The problem of fouling, that is, the dynamics of multiparticle deposition on porous surfaces, is analyzed by Kleinstreuer and Chin (1984). A comparison between *membrane* bioreactors is given by Kleinstreuer et al. (1984). Complications with the control of cell growth and substrate transfer resistance into the cell mass can also be expected. Hence, the achievement of a continuous, high-volume concentrated product stream is still a research goal. Experimental work with whole cells in hollow fiber reactors has been reported by Waterland et al. (1974), Kan and Shuler (1978), Matthiasson and Ramstorp (1981), and Roy et al. (1982). The results by Waterland et al. (1974) and Roy et al. (1982) were used to calibrate our computer simulation model discussed below. Roy et al. (1982) used an HFBR with different fiber packing in a batch recycle mode and a continuous flow apparatus for an analysis of lactic acid production and substrate uptake. The problems of excessive cell growth leading to fiber (bundle) contortion and diffusion limitations remain to be solved.

Huffman-Reichenbach and Harper (1982) reported problems with β-galactosidase retention and activity in hollow fiber bioreactors (polysulfone membranes) under different operational conditions. It is apparent that in many cases membranes no longer can be treated as simple mechanical sieves separating solutes according to different molecular weights. In general, it is important to consider the mode of reactor operation and the type of solute, solvent, and membrane, since enzyme retention characteristics depend in particular on the diffusion rate and convective flux, fluid shear stresses, axial flow rate, and enzyme characteristics (concentration, shape, size, deformability, etc.). Size of the "apparent" membrane pores and the corresponding enzyme retentions may be controlled by interactions of membrane–flux–flow–solvent–solute. For example, certain solvents may cause membrane swelling that alters the "apparent" pore size or enzyme shape and elasticity. In addition, radial volume flux may modify the membrane structure leading to solute leakage at higher flux values. Hence, in cases in which leakage is significant, a dynamic membrane theory has to replace the simple algebraic expressions for volume and mass flux across membranes based on irreversible thermodynamics (Kleinstreuer et al., 1984).

An early analysis for immobilized enzymes (Rony, 1971) was advanced by Waterland et al. (1974), Kim and Cooney (1976), and Webster and Shuler (1978). Webster et al. (1979) presented a solution for entrapped whole cells evenly distributed in the annular region. Recently, Webster and Shuler (1981) suggested a transient diffusion model to calculate the radial substrate concentration profiles in the spongy matrix. All these state-of-the-art models have certain features in common. The modeling equation is a reduced, linear form of the mass transport equation. Product kinetics are not considered, although under the assumption of a Gaden type I fermentation, the product concentration profiles could perhaps be deduced from substrate curves. The difference between the kinetics of free or pseudoimmobilized enzymes and entrapped whole cells is not addressed. Even if it is assumed that cell growth is terminated before the microorganisms are entrapped, the simplified Michaelis–Menten expression used is hardly applicable. The membrane functions as an ideal semipermeable barrier where concentration polarization and (radial) volume fluxes have been neglected. The simplified modeling equations are solved analytically.

A representative hollow fiber of which hundreds may be packed into one tube is schematically shown in Figure 3.14. A laminar feed stream in the lumen (region I) is a continuous source of substrate molecules that diffuse through the thin membrane (region II) into the stagnant annular region (spongy matrix region III) in which the microbial catalysts are well distributed. The governing equation for substrate transport and uptake in hollow fiber bioreactors can be deduced from Eq. (3.14) as follows:

$$\frac{\partial c}{\partial t} + v_z \frac{\partial c}{\partial z} + v_r \frac{\partial c}{\partial r} = \frac{D_r}{r} \frac{\partial}{\partial r}\left(r \frac{\partial c}{\partial r}\right) + D_z \frac{\partial^2 c}{\partial z^2} - kc^n \quad (3.45)$$

In existing analyses, $v_r \equiv 0$, $D_z \, \partial^2 c/\partial z^2 \approx 0$, and $n \equiv 1$. Kim and Cooney (1976) considered the steady-state version, whereas Webster and Shuler (1981) solved the reduced form of Eq. (3.45) without the convection term for the annular region. The sink term, kc, that is, biochemical conversion of the substrate, is a linearized expression of the Michaelis–Menten function, where $k = V_{max}/K_m$. An analytical solution of a simplified form of Eq. (3.45) with a parameter sensitivity analysis is presented by Kleinstreuer and Poweigha (1984). A comprehensive computer simulation model of the transient two-dimensional transport phenomena of an HFBR is given by Kleinstreuer and Agarwal (1986).

6. CONCLUSION

Stirred tanks operated in *batch* mode or intermittent feeding mode are commonly used in bioconversion processes; the exceptions include large single-cell protein plants and biotreatment of wastewater. *Continuous flow* stirred tanks with separation plus recycle and gas–liquid reactors have been commercialized for high volume production of valuable biomass such as single-cell protein. Fermentation plants for ethanol recycle both cells and spent beer from ethanol recovery. Immobilized or entrapped biocatalyst reactors including fluidized bed, packed bed, hollow fiber, membrane compartment, and microbial film bioreactors are still in the benchtop or pilot plant research stage. Full acceptance of the continuous flow mode and new reactor designs hinges on the following: (1) industry must be satisfied that large-scale process operation will be relatively trouble-free, (2) the new processes must be more profitable than existing processes, and (3) there must be a demand for products that existing processes cannot satisfy. It is evident that computer simulation can play an important role in achieving these objectives.

6. SYMBOLS AND NOMENCLATURE

6.1. General Nomenclature

$R_{P/x}$ Ratio of product unit to unit of biocatalyst

$\bar{y}$ Average growth yield

$Y_{P/S}$ Yield of feed (substrate) converted to final product

Y_{O_2} Oxygen yield

S_a Specific activity of biocatalysts

$\mathcal{V}$ (Reactor) volume

P Product concentration

X Concentration of active microorganisms

t Time

t_c Half-life of biocatalyst

F Volumetric flow rate

k_L Liquid side mass transfer coefficient

$k_L a$ Volumetric mass transfer coefficient

S Substrate concentration

a Effective mass transfer area (gas bubble surface area)

Q_p Rate of product formation

μ Specific growth rate coefficient

$\hat{\mu}$ Maximum growth rate coefficient

K_s Saturation constant

m Maintenance requirement, mass

Ne   Newton number (Eq. 3.5)

Re   Reynolds number (Eq. 3.6)

C_i Functions (cf. Bauer, 1979); $i = 1, 2, \ldots, 6$

ρ Density

∇ Grad operator

∇^2 Laplacian operator

$\mathbf{v}$ Velocity vector

c Concentration

S_c Sink or sources of c

$\mathbf{j}_c$ Mass flux

τ Stress tensor

$\mathbf{f}$ Body forces per unit volume

T Temperature

$\mathbf{q}_H$ Heat flux

π Total stress tensor

$\bar{H}$ Average enthalpy

r Reaction rate, radius

R Reaction term

$\dot{\gamma}$ Rate of deformation tensor

η Rheological coefficient, specific system property

p Pressure

δ Unit tensor

D Diffusion coefficient

I Interfacial mass transfer term

ϕ Gas volume fraction

G Arbitrary system property

D/Dt Substantial or Stokes' derivative

6.2. Airlift Fermenter

A_d Cross section of the downcomer

A_r Cross section of the riser

B Biomass concentration

B_A Dimensionless biomass concentration, $B/Y\,S_f$

c Oxygen concentration in the liquid phase

c^* Oxygen concentration in the liquid in equilibrium with air

D Dilution rate

D_1 Stoichiometric group $= Y_s S_f / Y_{ox} M_{ox}$ (dimensionless)

D_G Oxygen diffusivity in the gas phase

D_L Oxygen diffusivity in the liquid phase

F Volumetric flow rate

H Henry's constant

I_{LG} Interfacial mass transfer

J Average volumetric flux density of the mixture

J_G Gas volumetric flux density

J_L Liquid volumetric flux density

K Dimensionless Michaelis constant, K_m/S_f

K_m Michaelis constant

$k_L a$ Volumetric mass transfer coefficient

L Height of the riser

M_s Molecular weight of the substrate

M_{ox} Molecular weight of the oxygen

P_{tot} Total pressure

P_{O_2} Oxygen partial pressure

Q_L Liquid volumetric flow rate

Q_G Gas volumetric flow rate

R Reaction group, $L\mu/J_L$ (dimensionless)

r_B Rate of biomass generation

r_S Rate of substrate consumption

r_{O_2} Rate of oxygen consumption

S Substrate concentration

S_A Dimensionless substrate concentration, S/S_f

St_G Stanton number for the gas phase $(LK_L a/J_G)$ (RT/H) (dimensionless)

St_L Stanton number for the liquid phase, $LK_L a/J_L$ (dimensionless)

t Time

T Temperature

V Total volume of fermentor

V_G Superficial gas velocity

V_L Superficial liquid velocity

V_{SG} Gas-phase volume of the separator

V_{SL} Liquid-phase volume of the separator

Y_S Stoichiometric coefficient, biomass per substrate

Y_{ox} Stoichiometric coefficient, biomass per oxygen

Y_1 Dimensionless partial pressure of oxygen, $P_{O_2}/P_{O_2 0}$

Y_2 Dimensionless oxygen concentration, C/C^*

z Axial height

α Thermal diffusivity

β Ratio of separator volume to total volume

γ   Ratio of liquid-flow rates
φ   Gas hold up
η   Dimensionless tube length
μ   Specific metabolic rate
θ   Dimensionless dilution rate, DL/J_L
ξ   Ratio of cross section of riser to downcomer

6.3. Subscripts

d Related to the bottom of downcomer
f Related to the feed
G Gas phase
s Related to the gas separator
T Related to the top of the riser
o Related to the point of gas injection
L Liquid phase

6.4. Greek Symbols

α   Diffusion coefficient ratio
β   Hollow fiber radius ratio
η   Effectiveness factor
λ   Thiele modulus
Λ   Eigenvalues
σ   Overall mass transfer resistance

6.5. Input Data

T = 310 (K)
μ = 0.4 (hr)$^{-1}$
K_m = 37.5 (mg liter^{-1})
H = 900 (atm liter^{-1} mol^{-1})
Y_s = 0.4 (g) biomass per (g) substrate
L = 400–1000 (cm)
β = 0.1
J_G = 35 (cm sec^{-1})
S_f = 200–775 (mg liter^{-1})
ξ = 4.0
Y_{ox} = 1.0/(1.7/Y_s − 2.05) (g) biomass per (g) oxygen

6.6. Nomenclature for Webster and Shuler Model

C_o Inlet concentration

C_3/C_o Relative concentration

D_3 Effective diffusivity in spongy matrix

k First-order rate constant

K_m Michaelis constant

r/b Relative radius

b Inner radius of spongy matrix

t Time

V Maximum reaction rate

$\beta = d/b$ Relative radius of which there is no substrate concentration gradient

6.7. Input Data (Kim and Cooney, 1976)

6.7.1. Case Study (a) (C/C_o versus r all regions)

λ = 1.0, 10.0

β = 1.75

D_1/D_2 = 10.0

K_a = K_b = 1.0

α = 1

b/a = 1.005

Z = 0.05, 0.15, 0.25, 0.35, 0.45

6.7.2. Case Study (b) (η versus λ; parameter, K/α)

λ = 0.1–100.0

β = 1.37

Z = 0.13

K/α = 0.9, 1.0, 1.1

6.7.3. Case Study (c) (η versus λ; parameter, β)

λ = 0.1–100.0

β = 1.1, 1.37, 1.56

Z = 0.13, 0.25, 0.63

K/α = 1.0

REFERENCES

Aiba, S., Humphrey, A. E., and Millis, N. F. *Biochemical Engineering,* Tokyo, University of Tokyo Press, 2nd ed., 1973.

Anderson, D. A., Tannehill, J. C., and Pletcher, R. H. *Computational Fluid Mechanics and Heat Transfer,* New York, McGraw-Hill, 1984.

Andrew, G. F. *Biotechnol. Bioengr.* **24,** 2013, 1982.

Atkinson, B. *Biochemical Reactors,* London, Methuen, 1974.

Bailey, J. E. *Chem. Eng. Sci.* **35,** 1854, 1980.

Bailey, J. E. In Blanch, H. W., Papoutsakis, E. T., and Stephanopoulos, G. (eds.), *Foundations of Biochemical Engineering,* ACS Symp. Series 207, Washington, D.C., 1983.

Blenke, H. *Adv. Biochem. Eng.* **13,** 1979.

Bowski, L., Perley, C., and West, J. M. 182nd ACS Meeting, New York, 1981.

Box, G. E., and Jenkins, G. M. *Time Series Analysis: Forecasti and Control,* Oakland, CA, Holden-Day, 1976.

Brauer, H. *Adv. Biochem. Eng.* **13,** 1979.

Bravard, J. P., Coroonnier, M., Kerenevez, J. P., and Lebenault, S., *Biotechnol. Bioengr.* **21,** 1239, 1979.

Bravard, J. P. et al. *Biotech Bioeng.* **21,** 1239, 1979.

Brookey, R. S. *Chem. Eng. Commun.* **8,** 1, 1981.

Buchholz, K. In Fiechter, A. (ed.), *Adv. Biochem. Eng.,* NY, Berlin, Springer Verlag, Vol. 24, 1982.

Bungay, H. R., In Ratledge, C., Dawson, P., and Rattray, J. (eds.), *Am. Oil Chemists Soc.* 1984, pp. 45–54.

Cichy, P. T. et al. *Ind. Eng. Chem.* **61,** 6, 1969.

Cohen, A., Zoetemeyer, R. J., Van Deursen, A., and Van Andel, J. G. *Water Research* **13,** 571, 1979.

Cooney, C. L. *Science,* **219,** 728, 1983.

Dunn, I. J., and Mor, J. -R. *Biotechnol. Bioengr.* **17,** 1805, 1975.

Fan, L. -S., Long, T. -R., and Fujie, K. AIChE Annual Meeting, San Francisco, Nov. 25–30, 1984.

Guilbault, G. In Vieth et al. (eds.), *Biochemical Engineering,* New York, New York Academy of Sciences, 1981, vol. 326.

Goldberg, I., Rock, J. S., Ben-Bassat, M., and Mateles, R. I. *Biotechnol. Bioengr.* **18,** 1657, 1976.

Harder, A., and Roels, J. A., In Fiechte, A. (ed.), *Adv. in Biochem. Eng.,* Vol. 21, Berlin, Springer Verlag, 1982.

Hamer, G. *Biotechnol. Bioengr.* **24,** 511, 1982.

Hampel, W. *Adv. Biochem. Eng.* **13,** 1979.

Hatch, R. T. Ph.D. Thesis, M.I.T., Cambridge, 1973.

Hatch, R. T., and Wang, D. I. C. Presented at the 1st Chemical Congress of the North American Continent, Mexico City, December 1975.

Huffman-Reichenbach, L. M., and W. J. Harper, *J. Dairy Sci.* **65,** 887, 1982.

Kan, J. K., and Shuler, M. L., *Biotech. Bioeng.* **20,** 217, 1978.

Kim, S. S., and Cooney, D. D. *Chem. Eng. Sci.* **31,** 289, 1976.

Kiparissides, C., and MacGregor, M. Presented at the 182nd ACS Meeting, New York, Am. Chem. Soc., Aug. 23–28, 1981.

Kleinstreuer, C., and Poweigha, T. *Biotechnol. Bioengr.* **24,** 1941, 1982.

Kleinstreuer, C., and Poweigha, T. In Fiechter, A. (ed.), *Adv. Biochem./Biotechnol.* Berlin, Springer Verlag, vol. 30, 1984.

Kleinstreuer, C., and Chin, T. P. *Chem. Eng. Comm.* **28,** 193, 1984.

Kleinstreuer, C. et al. A Comparative Study of Membrane Bioreactors Presented at AIChE Annual Meeting, San Francisco, Nov. 25–30, 1984.

Kleinstreuer, C., and Agarwal, S. S., Biotech. Bioeng. **28**, 1233, 1986.

Kuraishi, M. et al. In *Microbial Growth on C1-Compounds,* Osaka, Soc. Ferment. Technol., 1975, p. 231.

Lam, H. Y. L., and Bungay, H. R. In Grayson, L. P., and Biedenbach, J. M. (eds.) *Engineering Images for the Future,* Proc. Conf. Am. Soc. Engr. 1983, ed. 2, p. 474.

LeGoff, P. *Chem. Eng. Sci.* **35**, 2029, 1980.

Leng, D. E. AIChE Meeting, San Francisco, 1981.

Luttmann, R. et al. *Biotechnol. Bioeng.* **24**, 817 1982a; **24**, 1851, 1982b.

Matthiasson, B., and Ramstorp, M. *Biotechnol. Lett.* **3**, 561, 1981.

Merchuk, J. C., and Stein, Y. *Biotechnol. Bioeng.* **23**, 1309, 1981 and *AIChE J.* **27**, (3) 1981.

Merchuk, J. C., Stein, Y., and Mateles, R. I. *Biotechnol. Bioeng.* **22**, 1189, 1980.

Messing, R. A. (ed.). *Immobilized Enzymes for Industrial Reactors,* New York, Academic Press, 1975.

Michaels, A. S. *Desalivation* **35**, 329, 1980.

Moo-Young, M., and Blanch, H. W. *Adv. Biochem. Eng.* **19**, 1981.

Nauman, E. B. *Chem. Engr. Sci.* **24**, 1461, 1969.

Nauman, E. B. *Chem. Engr. Comm.* **8**, 53, 1981.

Ollis, D. F. In Lapidus, L., and Amundson, N. R., *Chemical Reactor Theory,* Prentice-Hall, 1977.

Patterson, G. K. *Chem. Eng. Commun.* **8**, 25, 1981.

Pohland, F. G., and Ghosh, S. *Environ. Lett.* **1**, 4, 1971.

Poweigha, T., Theoretical Analysis and Computer Simulation of Biochemical Airlift Reactor Process Dynamics, Ph.D. Thesis, RPI, Troy, NY 1987

Prokop, A., In Blanch et al. (eds.), *Foundations of Biochemical Engineering,* Symp. Ser. No. 207, New York, Am. Chem. Soc., 1983.

Ramkrishna, D., Ghose, T. K. et al. (eds.), In *Advances in Biochemical Engineering,* Vol. 11, Berlin, Springer, 1979.

Rittmann, B. E., and McCarty, P. L. *Biotechnol. Bioeng.* **22**, 2343, 1980.

Roache, P. J. *Computational Fluid Dynamics,* Albuquerque, NM, Hermosa, 1972.

Rony, P. R. *Biotechnol. Bioengr.* **13**, 431, 1971.

Roy, T. B. V., Blanch, H. W., and Wilke, C. R. *Biotechnol Lett* **4**, 483, 1982.

Russell, T. W. F., Dunn, I. J., and Blanch, H. W. *Biotechnol. Bioengr.* **16**, 1261, 1974.

Schaftlein, R. W., and Russell, T. W. F. *Ind. Eng. Chem.* **60**, 12, 1968.

Schügerl, K. *Chem. Ing.-Tech.,* **52**, 951, 1980.

Schügerl, K. *Adv. Biochem. Eng.* **19**, 1981.

Schügerl, K. *Adv. Biochem. Eng.* **22**, 1982.

Schügerl, K. Lucke, J., and Oels, U. *Adv. Biochem. Eng.* **8**, 1977.

Shah, Y. T., Kelkar, B. G., Godbole, S. P., and Deckwer, W. -D., *AIChE J.* **28**, 353, 1982.

Shuler, M. L., and Domach, M. M. In Blanch et al. (eds.), *Foundations of Biochemical Engineering,* Am. Chem. Soc., 1983.

Sittig, W. *Chemtech,* 3, 606–613, 1983.

Stephanopoulos, G., and San, K-Y. Presented at the 182nd ACS Meeting, New York, Aug. 23–28, 1981.

Trevan, M. D. *Immobilized Enzymes.* New York, Wiley, 1980.

Vemuri, V. *Modeling of Complex Systems,* New York, Academic Press, 1978.

Vemuri, V., and Karplus, W. J., *Digital Computer Treatment of Partial Differential Equations,* Englewood Cliffs, NJ, Prentice Hall, 1981.

Wang, H. Y., Cooney, C. L., and Wang, D. I. C., *Biotechnol. Bioengr.* **21,** 975, 1979.

Waterland, L. R. et al. *AIChE J.* **20,** 1974.

Waterland, L. R., Robertson, C. P., and Michaels, A. S. *Chem. Eng. Commun.* **2,** 37, 1975.

Webster, I. A., and Shuler, M. L. *Biotechnol. Bioengr.* **20,** 1541, 1978; **23,** 447, 1981.

Webster, I. A., Shuler, M. L., and Rony, P. R., *Biotechnol. Bioengr.* **21,** 1725, 1979.

Zeigler, B. P., Meister, D., Dunn, I. J., Blanch, H. W. and Russell, T. W. F. *Biotechnol. Bioengr.* **19,** 507, 1977.

4

BIOMASS REFINING

GEORGE T. TSAO

Professor of Chemical Engineering and
Director of Laboratory of Renewable Resource Engineering
Purdue University
West Lafayette, Indiana

MICHAEL R. LADISCH

Professor of Agricultural and Chemical Engineering
Group Leader, Laboratory of Renewable Resource Engineering
Purdue University
West Lafayette, Indiana

HENRY R. BUNGAY

Professor of Chemical and Environmental Engineering
Rensselaer Polytechnic Institute
Troy, New York

1. GENERAL BACKGROUND

Although combustion, pyrolysis, and other thermochemical conversion routes to fuels or chemicals have their champions, bioconversion is fast approaching commercialization. There is little doubt that fermentable sugars from biomass will compete with traditional carbohydrates, but processes for obtaining these sugars need updated comparisons to account for continuing improvements. The underlying concept of biomass refining is low cost for carbohydrate polymers. Inexpensive lignocellulosic biomass contains cellulose, hemicellulose, and lignin, all of which can lead to salable products.

Biomass as an industrial feedstock has a long history, and many of the "new" processes draw heavily on past practice. For example, pulping with organic solvents is well known, and the new twist is to focus on lignin as a valuable coproduct. Steam explosion of wood chips was commercialized as the Masonite process in the 1930s, but awareness that the cellulose in exploded wood is easily hydrolyzed by enzymes is more recent.

Annual availability of low-valued cellulosic materials in the United States has been estimated to be about 1 billion tons (dry weight). This figure does not include high-valued cellulosic products such as pulp, lumber, and plywood. About 20% of the low-valued cellulosic materials can be collected at a reasonably low cost to support a chemical industry. The 200 million tons of biomass represents about 3.5×10^{15} Btu of energy. Assuming that 25% of this energy can be converted into that of, say, ethanol (excluding byproducts), a total of about 9 billion gallons of anhydrous ethanol can be produced to substitute for part of our gasoline fuel each year.

Cellulosic materials stand out for use as carbon and energy sources for production of chemicals because of great abundance. Unfortunately, great difficulties are encountered in achieving an efficient conversion. Pretreatment by some physical, chemical, and biological means is of key importance in unlocking the huge amounts of cellulosic renewable resources to assure a high process efficiency. Other raw materials from biological sources, such as cheese whey and starchy wastes, can be fermented readily by many types of microorganisms to generate various products in good yields. No pretreatment is necessary because starch, lactose, and soluble proteins can be readily metabolized by many living organisms.

Quality of the feedstock determines its use. Biomass rich in starch or sugars has great value as food, but most residues and wastes from agricultural operations have lower value as ground cover to prevent erosion or as a source of nutrient organic matter for the next growing period. There is high probability that inexpensive substitutes can be found for these applications so that such biomass can be used more profitably. Species of trees poorly suited for lumber or fiber may have greatest value as feedstocks. Many individual trees or crop plants are too deformed, diseased, or spoiled for their normal uses. Although the quantities of readily available biomass can match only a part of the enormous demand for fuels, there are copious supplies for the manufacture of chemicals. Success with the production of chemicals from biomass should provide investment capital and a better technological blueprint for developing other biomass industries.

1.1. Chemical Composition

General ranges of biomass compositions of several plants are given in Table 4.1. Cellulose is a linear homopolymer of anhydroglucose units linked by β-

TABLE 4.1 Chemical Composition of Cellulose Biomass

	Cellulose (%)	Hemicellulose (%)	Lignin (%)	Starch (%)
Hardwoods	40–55	24–40	18–25	—
Softwoods	45–50	25–35	25–35	—
Grasses; straws	25–40	25–50	10–30	—
Corn	3–4	2–3	1–2	75–85
Cotton	89	5	—	—
Cornstover	38	26	11	—
Tall fescue	34	26	8	—
Leaves	15–20	80–85	—	—
Newspaper	40–55	25–40	18–30	—
Birch	44.9	32.7	19.3	—
Spruce	46.1	24.6	26.3	—

Source: Compiled from Ladisch (1983) and Cowling and Kirk (1976).

D-1,4 glucosidic bonds. Having no side chains, cellulose molecules are sterically compatible for close association and form intramolecular as well as intermolecular hydrogen bonding. Cellulose is generally crystalline, with up to about 15% being "amorphous." The exact crystalline structure of cellulose is still a subject of debate among carbohydrate scientists. Pure starch is digestible by numerous living organisms. Pure cellulose is much more resistant; this is not necessarily because the β-glucosidic bonds are different from those of the α-configuration but rather because the crystalline structure makes cellulose difficult to hydrolyze not only in a thermodynamic sense but also because of reaction kinetics.

Hemicellulose is associated with cellulose and lignin and, supposedly, furnishes cell wall rigidity and flexibility. Hemicelulose is not a homopolymer but is rather complex in terms of both monomeric components and molecular structures. Generally speaking, hemicelluloses are a family of highly branched polymers. In hydrolysates of hemicellulose, a number of pentoses, hexoses, and uronic acids has been identified, including D-xylose, D-mannose, D-glucose, D-galactose, L-arabinose, D-glucuronic acid, 4-O-ethyl-D-glucuronic acid, and D-galacturonic acid (Whistler and Richards, 1970). Based upon their constituents, hemicelluloses have been classified into two groups, pentosan and hexosan, which exist in variable proportions in plants of different species. Hemicellulose, being highly branched, has been quite easy to hydrolyze by acids or enzymes. Warm water or mild aqueous alkaline solutions can also extract significant proportions of hemicellulose from native plant materials without much depolymerization. Many microbes can use the hemicellulose fraction, in either the native polymeric form or after hydrolysis. Thus, hemicellulose has little resistance to conversion into monomeric sugars.

Lignin is a polyphenolic macromolecule, and its chemical structure is not yet completely clear. The basic units of plant lignin are thought to be 3,5-

dimethoxy-4-hydroxyphenylpropane, 3-methoxy-4-hydroxyphenylpropane, and 4-hydroxyphenylpropane. The chemistry of lignin is of great interest to biomass conversion. Lignin is rich in carbon and hydrogen and has a fairly good fuel value. Some precursors of lignin such as paracumeryl alcohol, coniferyl alcohol, and syringyl alcohol would be valuable products. There are some processes that break lignins down into phenols, but the problem of separation is very challenging to say the least. A very complex mixture of soluble products results, and the degree of cross-linking of the lignins that are left in many cases will be higher than that of the starting material.

Extractives in biomass include waxes, fats, essential oils, tannins, resins, fatty acids, terpenes, alkaloids, and gums. There are various cytoplasmic constituents and nonextractives such as silica, alkali, alkali earth elements, and oxalates. In straws, the nonextractives may be up to 10% of the dry weight. As a consequence, the cellulose, hemicellulose, and lignin constituents of various biomass materials do not account for all of the dry weight of the biomass. The most significant differences occur for grasses (see Table 4.2).

1.2. Physical Structure

In addition to its crystallinity, several levels of architectural structure also exist in cellulosic materials and are pertinent to their conversion. Figure 4.1 shows the outline of a group of plant cells. The celluloses are the cell wall materials of plants. Surrounding the cell is the middle lamella, which is heavily lignified and contains lignin and hemicellulose in the proportion of approximately 7 to 3. Cellulose is contained in the cell wall. Within the cell wall, four distinct concentric morphological layers can be distinguished. The outermost layer is called the primary wall, which is about 0.1 mm thick and contains about one-third cellulose. In the primary wall, the microfibrillar structure appears as a loose and random network. The microfibrils are oriented around the fiber at approximately right angles to the axial fiber direction. The secondary wall forms during growth and maturation of the cell and contains three layers, an outer layer, S_1, a middle layer, S_2, and an inner layer, S_3. The microfibrils of the outer layers are wound in flat helices. The direction of winding alternates to form overlapping spirals. In the middle layers, S_2, the cellulose fibrils are tightly packed in a steep parallel helix. The innermost S_3 layer consists of helical microfibrils. The structure of this layer is not completely understood. For a raw cotton fiber, the primary wall is about 0.1–0.2 μm thick and the secondary wall is 1–4 μm thick and contains almost pure cellulose. Lignin, hemicellulose, and cellulose are in the primary wall. In general, cellulose occupies about 90% of the dry weight of the secondary wall in various fiber cells. The primary elements of supermolecular structures of cellulose are elementary microfibrils that are bound by linkage of individual linear polymeric cellulose molecules. X-ray diffraction and other evidence indicate that cellulose microfibrils are partially crys-

TABLE 4.2. Analysis of Various Biomasses[a]

Material	Hemicellulose							(1) Total Hemi-cellulose	(2) Cellulose	(3) Lignin	(4) Ash	Total (1) + (2) (3) + (4)
	Xylan	Araban	Glucan	Galactan	Mannan	Acetyl	Uronic Acid Anhydride					
Spruce	7.6	0.5	2.3	1.9	10.3	1.4	2.5	26.5	42.0	28.6	0.4	97.5
Pine	7.2	0.6	2.8	1.3	7.6	1.6	2.4	23.5	42.0	27.8	0.4	93.7
Birch	24.6	0.5	—	1.0	0.5	4.8	5.9	37.3	38.8	19.5	0.3	95.9
Balsam fir	4.6	0.5	1.9	1.0	11.8	1.4	3.2	24.4	42.7	28.0	0.2	95.3
Aspen	15.2	0.4	3.8	0.8	2.2	3.2	3.1	28.7	50.8	15.5	0.2	95.2
Red maple	16.4	0.5	5.3	0.6	3.3	3.6	3.3	33.0	39.0	23.0	0.2	95.2
Wheat straw[b]	18.5	1.6	2.2	0.7	0	2.4	2.2	27.6	34.0	18.0	1.3	81.0
Wheat straw	—	—	—	—	—	—	—	31.0	42.0	7.0	—	80.0
Cotton straw	—	—	—	—	—	—	—	12.0	42.0	15.0	—	69.0
Rye-grass straw	—	—	—	—	—	—	—	27.0	37.0	5.0	—	69.0
Cornstalk	—	—	—	—	—	—	—	26.0	38.0	11.0	3.0	78.0
Cornstalk[c]	18.0	2.2	3.6	0.8	0	3.6	4.4	32.6	33.5	11.0	1.0	78.1
Corn cob	25.0	3.0	—	—	—	—	—	28.0	36.5	—	—	83.0
Bagasse	—	—	—	—	—	—	—	34.0	38.0	11.0	—	83.0
Reed Canary grass	9.7	2.7	1.5	1.0	—	—	1.6	16.5[d]	28.0[e]	5.3[f]	—	49.4
Orchard grass	7.8	1.9	1.5	0.5	—	—	2.7	14.4[d]	30.0[e]	4.7[f]	—	49.1
Brome grass	8.7	2.4	1.7	0.5	—	—	1.7	15.3[d]	27.0[e]	5.5[f]	—	47.8
Tall fescue	15.6	3.7	1.4	1.0	—	—	2.5	24.5[d]	30.0[e]	3.1[f]	—	57.6
Reed foxtail	8.8	3.1	2.2	1.2	—	—	2.1	17.7[d]	29.0[e]	4.5[f]	—	51.2

Source: Ladisch et al., 1983.

[a]Whole biomass material, dry basis. Authors have attempted to present data on a comparable basis.
[b]Extractives assumed to be 18%, by difference—author's calculation.
[c]Extractives assumed to be 22%.
[d]By alditol acetate.
[e]By TFAA.
[f]By permanganate.

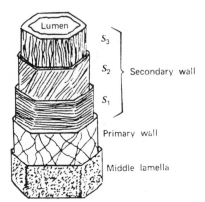

Figure 4.1 Architecture of a cell wall.

talline and partially disordered. The polymers are imbedded in a cell structure that is very tough and very hard to break up. Therefore, the grinding costs of reducing the particle size of untreated material can be very high. In biomass, lignin imbedded between cellulose microfibrils is indistinct and penetrates the outer parts of the cellulose structure with hemicellulose. Such structure is very difficult to see even with the most powerful electron microscope. Lignin in this matrix is different from the lignin outside.

When cows digest forage, the microorganisms in the ruminant stomach attack the inside of the cell first, the so-called lumen. They would have a very difficult time attacking from the outside of the cell. This may also be true for a direct fermentation of cellulose to ethanol using *Clostridium thermocellum* or *Clostridium saccharolyticum* because these anaerobes are very similar to organisms in the rumen system. Although there is much literature on rumen microbiology having to do with digestion of cellulose, the fundamentals of cellulose utilization are still not clearly understood.

For a better concept of microfibril structure, see Figure 4.2, which is based on electron micrographs by Delawig. Each box stands for many cellulose polymers crystallized together with a certain order. In the cross section shown in the figure, lignin surrounds the materials whereas hemicellulose penetrates the lignin. Furthermore, the material has a very organized structure. It is about 30 Å on a side, and four line up to give a 120 Å width. Not every plant material will have a similar structure, but the structure is very organized and very crystalline. Although lignin is often referred to as a polymer, it is really a three-dimensional matrix because there is no specific repeat unit. Consequently, this material is very gluelike.

The above mentioned structural features are pertinent to conversion of cellulosics. In enzymatic hydrolysis, the prerequisite for a reaction to take place is a direct association between the molecules of the enzyme and the substrate. In this case, the barrier represented by the lignin-rich middle lamella makes the diffusion and the penetration by the macromolecular enzymes extremely difficult and slow, if not impossible. The phenolic groups in lignin may even be inhibitory toward enzymes.

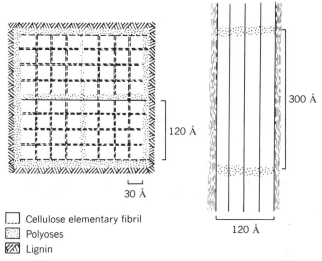

Cellulose elementary fibril
Polyoses
Lignin

Figure 4.2. Microfibril. (From Fengel, 1971.)

The lignin layer is expected to cause less of a barrier for acid hydrolysis because of the relatively small molecular size of the acid catalysts. This size difference between the acid catalysts and enzymes might explain the basic difference in the effectiveness of the two types of hydrolysis catalysts. Enzymes are expensive, difficult to diffuse, and slow in attacking an insoluble and highly crystalline substrate. Acids are less expensive and fast in diffusion. The main resistance to acid hydrolysis is the cellulose crystalline structure. Acids may cause problems in the formation of by-products such as furfural under the low pH and high temperature conditions of cellulose hydrolysis.

Enzymatic hydrolysis of cellulose is slow because of the inherent difficulty in the direct contact between a macromolecule of enzyme and a highly crystalline substrate. Fermentation processes for producing the cellulase enzymes are already very efficient. When cellulases are used to attack soluble cello-oligomers in the laboratory, the hydrolysis rate is quite fast indicating, indeed, that the β-1,4-glucosidic linkage is not particularly difficult to hydrolyze. What makes the rate of enzymatic cellulose hydrolysis 10–100 times slower than the corresponding rate of enzymatic hydrolysis of starch is not the difference in the glucosidic linkages in the two types of polyglucose molecules but rather the cellulose being insoluble and crystalline and starch being easily hydrated and gelatinized. Starch hydrolysis by α-amylase and glucoamylase, which is a well-established industrial process, requires an enzyme cost of about 1 cent per pound of glucose formed. If 10–100 times as much enzymes are used in cellulose hydrolysis to achieve a comparable rate to that of starch hydrolysis, the cost of cellulases will be prohibitively high if the enzyme is used only once and discarded. A major research breakthrough on techniques of recycling cellulases before enzymatic cellulose hydrolysis can be competitive may be necessary.

2. ACID HYDROLYSIS

Acid hydrolysis of cellulosics can be divided into two groups: high temperature and low temperature. The low-temperature acid process ranges from 80 to 140°C; most high-temperature processes operate from 160 to 240°C. The main difference due to temperature in the reaction is the formation of furfural and its derivatives at a high temperature in the acidic solution. Furfural and related by-products represent a loss of sugar yield and create problems in subsequent fermentations because of inhibitory effects of these by-products on living cells. The acid is also much more corrosive at a high temperature, and the special alloys for equipment fabrication are costly.

Concentrated mineral acids including phosphoric acid (70% strength or more), hydrochloric acid (40%), sulfuric acid (62% or more), and aqueous zinc chloride (a Lewis acid, 72% or more) are known to be able to swell and dissolve cellulose. After pretreatment by a concentrated mineral acid, the decrystallized cellulose can be readily hydrolyzed. Generally, a concentrated mineral acid acts more as a "solvent" for cellulose dissolution than as an acid catalyst for hydrolysis. After addition of water, the diluted acid will become an effective catalyst for cellulose hydrolysis. The low-temperature acid process involves the use of a mineral acid in concentrated form first to decrystalize and then after dilution with water to hydrolyze cellulose.

Another version of a dilute acid process of cellulose hydrolysis involves heating the cellulosics at a high temperature in the presence of water. Under these conditions, organic acids, mostly acetic acid, will be released from the plant materials. The released acid can then promote additional hydrolysis. This is a mild acid hydrolysis with little or no external addition of acid, and thus it has been known as the autohydrolysis.

Figure 4.3 shows different cellulose models. Some are folded chains and some extend, and disputes about the various models continue. However, there is enough known for certain about cellulose structure for use in an engineering context. Basically, the argument is between the fringed micell type model and the folding chain model sometimes called Ellefson's fibrillized model, also referred to as the folding chain model. Incidentally, the seven lines have meaning because seven linear sections are in each little square in Figure 4.2. The extended model has each of these lines as a polymer made up of anhydrose glucose units, and there are amorphous regions or disorder. The folding chain model, instead of amorphous regions, has one basic polymer with a molecular weight corresponding to a degree of polymerization (DP) of about 1400 glucose units. There are seven repeat units and the amorphous regions are the folds that are strained a little and are susceptible to rupture.

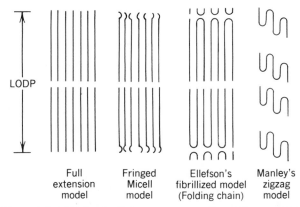

LODP

| Full extension model | Fringed Micell model | Ellefson's fibrillized model (Folding chain) | Manley's zigzag model |

Figure 4.3. Models of cellulose structures. (From Chang, 1971.)

3. SOME DETAILS OF A LOW-TEMPERATURE ACID HYDROLYSIS PROCESS

3.1. Hemicellulose Hydrolysis

Hemicellulose can be hydrolyzed with 5–10% sulfuric acid at 100°C, dilute acid (0.1–0.4% sulfuric acid) at 140 to 170°C, or strong acid (30–60%) at low temperatures (30–80°C). For example, corn residue (stalks) are hydrolyzed with 5% sulfuric acid at 100°C with the hydrolysis carried out at 50–80% moisture for 4–6 hr. The sugars are recovered by leaching the hydrolyzed material. A 10–15% sugar solution can be obtained using an appropriate leaching scheme. Xylose can be dehydrated using acid to form furfural, a chemical now commercially derived from hemicellulose. Since hemicellulose can account for about 30% of the biomass, its use is essential for a profitable overall biomass conversion plant. In this context, the fermentation of xylose to ethanol, either using glucose isomerase (EC5.3.1.5) to convert xylose to xylulose, which is then fermented to ethanol, or direct xylose fermentation, is essential. Improvements in yields (currently about 60%) and ethanol concentration (less than 5%) are required for fermentation of xylose to ethanol to be commercially successful. Arabinose is also present in significant quantity in the hydrolyzate of hemicellulose from agricultural residues (Voloch et al., 1984).

Cellulosic materials are generally low in bulk density. A sufficient amount of water to submerge a pile of wood chips or cornstalks in a container gives a minimum liquid to solid ratio of 5:1. After hydrolysis, even at 100% sugar yield, the obtainable sugar concentration in the hydrolysate will be low. The well-known Madison acid hydrolysis, for example, gives a solution of less than 5% by weight of total reducing sugars. To overcome this dilution problem, a "roasting–leaching" technique can be used to achieve a high concentration of xylose and other soluble sugars in the hemicellulose hydrolysate.

For most farm residues such as cornstalks, the moisture content can be anywhere from 15 to 50% when collected and delivered to the plant site. Green wood chips usually contain about 50% moisture. The first step of the roasting–leaching processing is to add a dilute acid to the biomass and uniformly distribute the acid in biomass. For agricultural residues, spraying a predetermined amount of sulfuric acid on the cellulosics solids can be easily done. A gentle rotating action will help distribute the acid.

The acidified plant biomass containing 50% moisture will appear "dry" and feel "dry" to the touch; it does not contain enough moisture to show any free liquid. This material is then heated with either live steam or some type of waste flue gas to 80–100°C. The sulfuric acid is not very corrosive at this temperature range. With no freely flowing liquid, the contact between the acidic wet solids and the container inner wall is also at a minimum.

After the 80–100°C treatment for several hours, 90% of the hemicellulose is solubilized. However, the monomeric and oligomeric products are still inside the cavities and pores in the bulky biomass as there is only a limited amount of moisture in the system at this point. The leaching step that follows will extract the solubles out with warm water.

Low temperature is desirable because at 80–100°C, xylose, arabinose, and uronic acids in the reaction mixture are stable for at least 24 hr without detectable furfural formation. According to published literature, in order to reduce furfural formation from acidified pentose solutions, we should use a high temperature but a short reaction time to carry out the hydrolysis. This is apparently true for the high-temperature range of 180°C or so, judging from the published values of the activation energy and kinetics constants of the involved chemical reactions. However, that preference apparently cannot be extrapolated to the low-temperature range of 80–100°C.

The solubilized hemicellulose carbohydrate in the internal cavities and pores of the plant biomass can be extracted with added warm water at 60–100°C. Depending upon the level of oligomers in the mixture, a slightly higher temperature will finish the acid-promoted hydrolysis to monomers once an extra amount of hot water is added. The roasting–leaching technique will allow the production of a hemicellulose hydrolysate of a relatively high concentration of solubles. For example, a typical plant biomass with about 35% by weight hemicellulose and a reaction mixture of 50% moisture results in a concentration of solubles after hydrolysis of about

$$100 \times \frac{35 \text{ lb solubles}}{100 \text{ lb moisture} + 35 \text{ lb solubles}} = 26\%$$

When hot water extracts the solubles, dilution is inevitable. However, by operating the leaching in a plug-flow trickling bed, the first portion of the exiting effluent will be high in sugar concentration. As additional volumes of effluent are collected, the sugar concentration will be progressively lowered.

Two cuts are taken: one above a preselected cumulative concentration of about 15% by weight of dissolved solids and one below this concentration. The weak juice can be recycled to extract sugars in the next column. Thus, by the roasting–leaching techniques, we can obtain a fairly strong hemicellulose hydrolysate for subsequent processing.

Syrups composed of mixed sugars from hemicellulose are available. The Masonite Company added a washing step to their explosion process and sold a concentrated syrup for cattle feeding. It was not particularly good in comparison to molasses and was not marketed aggressively. Nevertheless, similar syrups could be produced on short notice if there were sufficient demand. There are also waste streams from paper pulping that are rich in sugars but are contaminated with many ill-defined substances. Such streams also contain some acetate.

3.2. Cellulose Hydrolysis

3.2.1. Pretreatment

Cellulose in the lignocellulosic solid residue, after roasting and leaching to remove hemicellulose, will be hydrolyzed by a dilute acid to yield glucose with or without a prior ethanol extraction to isolate oligomeric lignin. As stated before, the presence or absence of lignin does not greatly affect the hydrolysis catalyzed by an acid but lignin may impair enzymatic hydrolysis.

After hemicellulose removal, the physical integrity of the solid residue is greatly reduced. The lignocellulose solids can easily be crushed to very fine powders with little power input. At this point, however, the crystalline structure of the cellulose remains, essentially, intact. The cellulose crystallinity could even be increased beyond that of the native biomass because of the possible, partial selective removal of the "amorphous" cellulose by the roasting–leaching treatment. In order to achieve a high level of cellulose conversion, a pretreatment of some sort is necessary.

The lignocellulose residue crushed to fine powder can be dewatered to 40–45% moisture by pressing. Before filtering and pressing, adding fresh acid will adjust the acid strength to a predetermined level. The wet cake by then has a substantial amount of acid uniformly distributed throughout. The wet cake is gently heated to remove water by evaporation and thus gradually increase the acid strength in the solid to decrystallize cellulose in a uniform manner. A suitable combination of the level of drying, the particle size, the amount of acid, the drying temperature, and the length of drying time will allow an optimal level of cellulose decrystallization by this pretreatment.

3.2.2. Cellulose Hydrolysis after Pretreatment

A sufficient amount of water will be added to the wet mixture described above to reduce acid strength in the liquid. At the diluted level, the acid becomes more of a catalyst for hydrolysis. The slurry is heated at 80–140°C to produce glucose. The effectiveness of the pretreatment, the temperature, and the time of reaction will determine the glucose yield. Alternatively, we may want to leave a sufficient amount of unreacted cellulose to be used together with lignin as the boiler fuel for the whole plant. In this case, the targeted level of desired cellulose conversion will dictate the temperature and the time of the hydrolysis step, as well as the extent of the pretreatment. Generally speaking, a very high level of cellulose conversion into glucose will require a disproportionately high level of pretreatment and thus the addition of a large amount of concentrated sulfuric acid. This situation of diminishing return by the applied solvent can be quantified in an economical study, and the optimal levels of pretreatment and hydrolysis can, we hope, be determined. In some cases, perhaps we would want to convert all the cellulose into glucose and would prefer to use a different fuel for steam generation.

Glucose is much more stable than xylose when heated in the presence of an acid. Furthermore, because of the decrystallization pretreatment that enhances cellulose reactivity, subsequent cellulose hydrolysis can be carried out at the relatively low temperature of 80–140°C. Therefore, the usual problem of decomposition of glucose to form hydroxymethyl furfural and other undesirable by-products can be minimized in this process.

The hydrolysate can easily be separated from residual cellulose and lignin by filtration and cake washing. The cake rich in lignin no longer holds much water. The acidic glucose solution can be neutralized and then fermented to chemical products. A process option shown in Figure 4.4 is combining glucose from cellulose with xylose and other sugars from hemicellulose for fermentation or other use. In this case, the acidic glucose solution from cellulose hydrolysis is used to impregnate, roast, and leach the original cellulosic raw material to hydrolyze hemicellulose. Even though this would mean a prolonged exposure of glucose to an acidic condition at a moderate temperature of 80–100°C (roasting temperature) for several additional hours, there is no serious problem of glucose decomposition because glucose is stable under those conditions. In this process option, the same acid is used three times: once in the concentrated form to crystallize cellulose, once after dilution in cellulose hydrolysis as a catalyst, and once as a catalyst for hemicellulose hydrolysis (Tsao et al., 1982).

Alternative processes do not handle thick solids, and cellulose hydrolysis in slurry reactors results in a low concentration of sugars. Consequently, subsequent separation steps become very energy intensive. The upper concentration limit is set by the bulk property of the biomass rather than by any of the inherent kinetics. A drop in yield results with increasing temperatures because the constant for hydrolysis changes more slowly than does the constant for degradation. Wenzl (1970) provides an excellent overview of producing sugars.

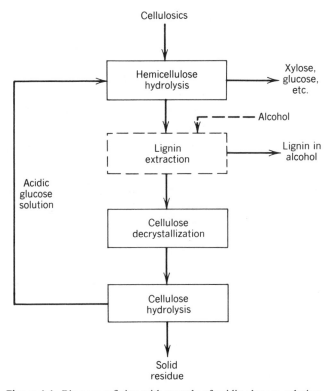

Figure 4.4. Biomass refining with recycle of acidic glucose solution.

Initially, the DP will drop off very quickly with a 2–4% weight loss in the cellulose structure to produce very little soluble products. For a long time, degree of polymerization will stay essentially constant even though sugars are being released. This is the so-called LODP, or leveling off degree of polymerization. This can be explained by either model as acid and water can penetrate the regions of disorder in the cellulose. Consequently, in a very short period of time, these regions will be hydrolyzed away. What remains corresponds to the LODP. For example, with a polymer of DP of 1400, acid hydrolysis for a short period of time by the folding chain model gives seven fragments (7 divided into 1400 is 200). The LODP for different materials is approximately 200. Pretreatment that increases the proportion of the amorphous to crystalline regions decreases LODP. Now acid and water can penetrate more of the cellulose structure so the easy part of the cellulose for hydrolysis is larger and therefore the crystalline structure that is left has a much shorter LODP. In terms of acid hydrolysis, for pretreatment to be useful, a large reduction in the LODP is desired. This is not necessarily true for enzyme hydrolysis.

4. USE OF HEMICELLULOSE HYDROLYSATE

The single factor that will have a decisive effect on the overall process economics in conversion of cellulosics is the use of hemicellulose hydrolysate. A typical chemical composition of North American hardwood is shown in Table 4.3; the total hemicellulose consisted of pentosans, hexosans, uronics, and acetyls, which is about 35% of the total biomass weight. Among the various components, xylan, the polymer of anhydroxylose is the most prominent. Xylose was considered nonfermentable, meaning that the common yeast, *Saccharomyces,* could not ferment it to ethanol. Recently it was discovered that xylulose, an isomer of xylose, could be readily fermented to yield ethanol, and xylose could be converted into xylulose by a treatment with an enzyme called xylose isomerase, also known as glucose isomerase. We will examine four options for hemicellulose use.

4.1. Production of Furfural

Pentoses upon heating in the presence of an acid lose water and form furfural.

$$C_5H_{10}O_5 \xrightarrow{-3H_2O} C_5H_4O_2 \longrightarrow \text{Decomposition products} \qquad (4.1)$$

$$\text{Pentose} \qquad\qquad \text{Furfural}$$

$$
\begin{array}{c}
\text{CHO} \\
| \\
(\text{HCOH})_4 \\
| \\
\text{COOH}
\end{array}
\xrightarrow[\;\searrow CO_2\;]{\text{heat}}
\begin{array}{c}
\text{CHO} \\
| \\
(\text{HCOH})_4 \\
| \\
\text{H}
\end{array}
\xrightarrow{-3H_2O} \text{Furfural} \qquad (4.2)
$$

Uronic acids   Pentoses

Uronic acids upon decarboxylation by heat will yield pentoses and, thus, are also potential raw materials for furfural formation. The theoretical maximum yield is 64 lb of furfural per 100 lb of pentoses consumed.

Furfural is a chemical of considerable uses as a solvent and also as a feedstock. The very first nylon was, in fact, made from furfural. Furfural is currently sold at over 50 cents per pound, a price that discourages its use. By the technique of roasting and leaching, a fairly concentrated acidic solution of pentoses and uronic acids can be produced relatively inexpensively. In this processing sequence, there are relatively few expensive costs before the concentrated hemicellulose hydrolysate. In fact, with the acid already in the hydrolysate, we can convert pentoses and uronic acids into furfural by simply heating it to about 260°C in a tubular reactor. As shown in Eq. (4.1), furfural under the acidic, high-temperature condition can form decomposition products of unknown identity. Simple heating with no special precau-

TABLE 4.3. Chemical Composition of Wood

Components	Weight Percent		
Cellulose			45.0
Hemicellulose			34.5
Pentosans		19.4	
Xylan	18.9		
Araban	0.5		
Hexosans		7.1	
Glucan	3.9		
Galactan	0.8		
Mannan	2.4		
Uronics		4.1	
Acetyl		3.9	
Lignin			20.2
Ash			0.3

tion will give a furfural yield of about 65% of the theoretical maximum, that is, about $64 \times 0.65 = 42$ lb of furfural per 100 lb of xylose equivalent (PE). Since xylose, arabinose, and uronic acids can all be converted into furfural as shown in Eqs. (4.1) and (4.2), the term PE is used to express the total available raw material for furfural formation. With a special-process precaution, for example, removing furfural from the aqueous solution as soon as it is formed to prevent its decomposition, 90% of the theoretical maximum yield is possible.

Furfural, besides being useful as a chemical and a solvent, is also an antiknock agent for improving octane ratings of gasoline. However, since furfural is an unsaturated aldehyde, it has a tendency to form a gummy polymeric mess under engine combustion conditions. By a selective hydrogenation, furfural can be converted into furfuryl alcohol, which is also a good antiknock agent, as well as having other uses.

4.2. Production of Single-Cell Protein as a Feed By-Product

The common *Saccharomyces* and *Candida* can grow very fast using xylose, arabinose, uronic acids, and acetic acid as the carbon source to yield new yeast cells. The cells of *Saccharomyces* can be used as the seed for ethanol fermentation of glucose in cellulose hydrolysate, and the yeast cells can also be harvested, dried, and marketed as a feed by-product. In this case, practically all the available carbon source in the hemicellulose hydrolysate is used, and thus subsequent wastewater treatment is minimized. For a small-scale operation, this approach of hemicellulose use could be the simplest in terms of process design and product marketing while meeting the requirements of environmental regulatory agencies.

4.3. Production of 2,3-Butanediol and Methyl Ethyl Ketone

A bacterial culture, *Klebsiella oxytoca,* can use xylose, arabinose, and also uronic acids under anaerobic conditions. From xylose and arabinose, the bacterial cells can produce 2,3-butanediol in a high yield (90% of the theoretical maximum) in a fairly high concentration of 9% by weight in the final fermentation broth. This diol has a number of possible uses such as a good antifreeze agent. This fermentation product, however, is difficult to purify by distillation because its boiling point is 180°C. A method was developed for converting 2,3-butanediol in the fermentation broth directly into methyl ethyl ketone (MEK) by acidic dehydration in nearly a quantitative yield. Methyl ethyl ketone, which boils at about 78°C, can be easily purified by distillation and has a well-established market.

4.4. Ethanol from Xylose after Isomerization

After a great deal of research effort, we can now convert xylose into ethanol. Other components including arabinose, acetic acid, and uronic acids cannot yet be used for this purpose. Currently, there are two parallel approaches of research and technology development dealing with the problem of xylose conversion into ethanol. One will be covered here; the second approach, which is based upon gene splicing work, is advancing rapidly. The conversion is shown briefly in Figure 4.5. Xylulose can be fermented to ethanol via the biochemical metabolic reaction sequence known as the pentose phosphate cycle. Xylose cannot be converted to ethanol unless it is first converted into xylulose. The common *Saccharomyces* yeast lacks the enzyme xylose isomerase, which is responsible for the conversion of xylose to xylulose and thus cannot ferment xylose to ethanol. The enzyme, xylose isomerase, also known as glucose isomerase, happens to be the commercially available enzyme extensively used in the industrial production of high fructose corn syrup (HFCS). Therefore, one approach for ethanol production from xylose is to conduct a simultaneous enzyme reaction and yeast fermentation. *Saccharomyces* yeast with the help of externally added glucose isomerase will convert xylose into ethanol. The yield of this process option is about 80% of the theoretical maximum; that is, about 0.4 lb of ethanol can be produced from each pound of xylose. However, the external addition of an industrial enzyme designed for a food product, HFCS, adds expense to ethanol production. There are several large corn-processing companies in the United States that apply this enzyme for producing huge amounts of HFCS. The usual industrial practice is to run an immobilized enzyme complex for only three half-lives. As the definition of half-life suggests, at the end of the three half-lives, the enzyme activity becomes $(\frac{1}{2})^3 = \frac{1}{8}$ of that of the fresh enzyme complex. However, since xylose is the natural substrate of this enzyme (xylose isomerase is a better name than the common commercial name glucose isomerase), the isomerase is much more active (about

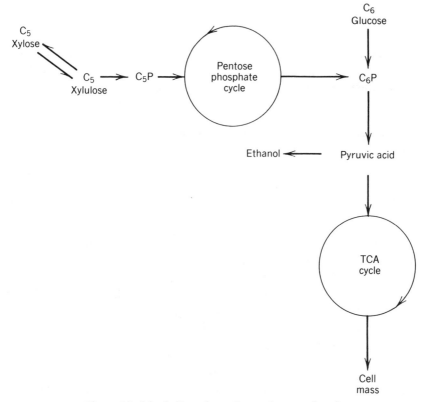

Figure 4.5. Metabolic pathway from xylose to ethanol.

three times more) toward xylose than glucose. Therefore, even the spent enzyme after three half-lives in HFCS production can still be very active for xylose conversion into xylulose. For companies with large HFCS operations, the spent glucose isomerase that otherwise means a cost of solid waste disposal could be used for ethanol production via this first process option. This approach is likely to be too costly for companies that have to purchase fresh enzyme specifically for ethanol production from xylose.

5. ENZYMATIC HYDROLYSIS OF CELLULOSE

Enzymes are a class of proteins that function as reaction catalysts in biological systems. Cellulose biodegradation is a natural phenomenon, otherwise the billions of tons of fallen trees and other green plants would accumulate each year and would eventually stop the life cycles on Earth completely. Pioneering work on enzymes that are responsible for cellulose hydrolysis and known as cellulases has generally been credited to a group of scientists at the U.S. Army Natick Laboratory. Reese et al. (1950) of the Natick group first suggested that two enzyme components were involved in the hydrolysis of cellulose. One designated as C_1 hydrolyzes native cellulose into short chains. The other, C_x, carries out the solubilization of short chains into sugar molecules. Since then, much research throughout the world has elucidated the mechanism of cellulolysis, which is a subject of great interest to plant pathology, animal husbandry, carbohydrate chemistry and enzymology, and, more recently, renewable resources and biomass use.

Nowadays, most agree that there are three major enzyme components involved in the hydrolysis of cellulose to generate small sugar molecules: (1) β-glucosidase (also known as cellobiose), (2) 1,4-β-glucan cellobiohydrolase (also known as C_1, avicellase, cellobiohydrolase, and exoglucanase), and (3) 1,4-β-glucan glucanohydrolase (also known as C_x, endoglucanase, and CMCase).

C_x is considered the enzyme component that promotes the initial attack on insoluble cellulose substrate by cleaving glucosidic linkages in a "random" manner. C_1 produces mostly cellobiose. Its primary function is to cleave off cellobiose units one at a time from a shrinking glucosidic chain. Cellobiase is the third major enzyme component that hydrolyzes cellobiose to glucose and thus completes the overall process of conversion of insoluble cellulose to glucose. However, the real situation is more complicated. For instance, the enzyme, cellobiase, is strongly inhibited by its product, glucose. When cellulose is hydrolyzed by a mixture of the three enzymes in a stirred-tank reactor, as many have attempted to do recently, there will be substantial amounts of cellobiose accumulated together with glucose. In a natural environment in which glucose, as soon as it is formed, will be metabolized by some living system, a complete degradation of cellulose without any substantial accumulation of glucose or cellobiose is often the case.

6. PERSPECTIVE

Processes for alcohol fuels from cellulose have had sufficient research, development, and cost analysis to show that complicated technology for hydrolyzing cellulose will not be profitable if hemicellulose and lignin are thrown away. When over half of the feedstock is wasted or used to defray the cost of boiler fuel, the glucose from cellulose must bear all the plant overhead. Fortunately, there are some fine potential uses for hemicellulose and lignin.

Although alcohol from cellulose is technically feasible, it probably is not yet competitive with older processes. Comparison of acid and enzymatic hydrolysis showed each to have advantages and disadvantages, and acid

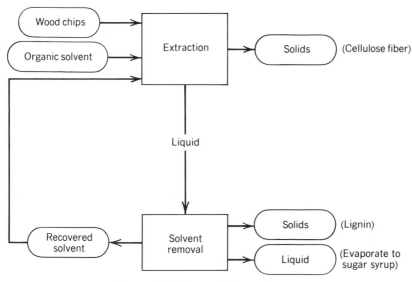

Figure 4.6. Solvent refining of biomass.

hydrolysis may be more economical. The conclusion was that it is premature to decide now, and several processes deserve more research and development.

Solvent refining is shown in Figure 4.6. Wood chips are extracted with a solvent containing some water with acid or base. Suitable solvents include butanol, ethanol, phenol, and formic acid. While lignin is dissolved, hemicellulose is hydrolyzed to sugars. The liquid is removed and treated to yield lignin as a solid or resinous material and a solution of sugars that is concentrated and used for animal feeding or as a fermentation substrate. The cellulose residue is fibrous and may be used in paper or may be a feedstock for cellulose derivatives. Rough estimates of values based on starting with 1 ton of wood chips (dry basis) at $30–50/ton are

Lignin	.2 ton	$300–600/ton
Fiber	.45 ton	$250–450/ton
Syrup	.2 ton	$30–90/ton

There is good potential profit if the cellulose serves as fiber, but costs of using solvents are too high if the cellulose is considered only as a source of glucose for fermentation. Note that there are no biological steps in this scheme.

Steps in a steam explosion process are shown in Figure 4.7. Wood chips are impregnated with high-pressure steam, and sudden release of pressure shatters the cellular structure. The conditions hydrolyze all of the hemicel-

George T. Tsao, Michael R. Ladisch, and Henry R. Bungay

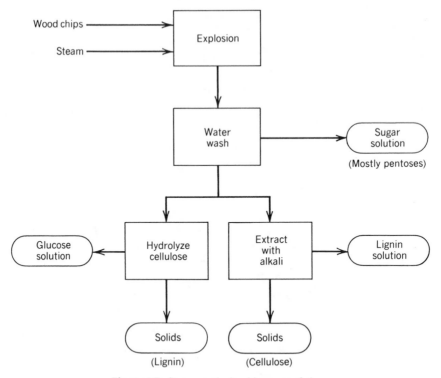

Figure 4.7. Steam explosion biomass refining.

lulose and degrade some of the resulting sugars. The next step should be a water wash to remove these sugars and various impurities and inhibitors of subsequent bioprocessing steps. As with solvent processing, these extracts can be concentrated to a syrup, but their value is somewhat diminished by the substances formed during steam explosion. Lignin extracted from the solid residue is equivalent to lignin from solvent processing. The cellulose remaining after lignin removal is over 90% pure, but the fibers are too weak for use in paper and the degree of polymerization is too low for making most cellulose derivatives. This cellulose can be considered as a cheap filler material or as a feedstock for bioconversion. The main advantage over solvent processing is lower cost. Feedstock and product values are the same as before except that the cellulose is worth $60–100/ton as a source of sugars. Two alternatives are shown in Figure 4.7, hydrolysis with lignin present or with lignin removed.

Direct fermentation of cellulosic biomass is shown in Figure 4.8 The preferred feedstock at present is cornstover because it is easily shredded to an acceptable particle size and is rich in carbohydrates. Note that all of the lignin is lost in the spent solids. The products are ethanol and boiler fuel.

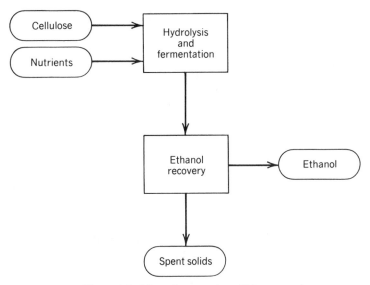

Figure 4.8. Direct fermentation of biomass.

This appears to be an inferior process because the product values are low. However, as processing is quite simple and costs should be very low. As a route to ethanol, this method is much more attractive than using costly steps for producing enzymes, for saccharification, and for fermentation to ethanol as in Figure 4.8. In other words, one step encompasses all of these operations.

There are good reasons for handling the sugars from hemicellulose separately. The best available organisms for ethanol fermentation tend not to accept the pentose that predominates in this fraction. Furthermore, organisms that can ferment pentoses and glucose together use the glucose first. Tolerance to ethanol is not very good for these organisms, so early accumulation of ethanol formed from glucose slows the rate of fermentaton of the other sugars.

Direct fermentation of partly purified cellulose from a steam explosion process should perform well. A preliminary test using strains developed at Massachusetts Insitute of Technology showed promise, but the cellulose was not completely used (Wang, D.I.C., unpublished). This is characteristic of direct fermentation processes, but strains having enzymes more active for this weakened cellulose can probably be found. When a direct fermentation of the cellulose to ethanol is perfected, refining based on steam explosion should be very attractive for any nation with biomass resources (Bungay, 1983). The syrups for cattle feed aid the agricultural sector, the lignin can be invaluable for advancing new construction methods based on less expensive adhesives for chipboard, and the ethanol will impact on needs for liquid fuel.

7. PRELIMINARY COST ESTIMATES

In a biomass conversion process, the major costs, exclusive of capital, are substrate cost (S_{cost}), pretreatment cost (P_{cost}), and hydrolysis cost (E_{cost}). These parameters are given by Ladisch et al. (1983):

$$S_{cost} = \frac{C_s}{X_{cell}\, Y_{GE}} \qquad (4.3)$$

$$P_{cost} = \frac{C_p}{X_{cell}\, Y_{GE}} \qquad (4.4)$$

$$E_{cost} = \frac{W_{cat} C_E}{G_E} \qquad (4.5)$$

where C_s, C_p, C_E are unit substrate, pretreatment, and enzyme costs: W_{cat} is the amount of catalyst (enzyme) used per liter of hydrolysis volume; G_E is the glucose (product) concentration in grams per liter; and Y_{GE} is the fractional glucose yield from cellulose. For example, let C_s = \$0.03/lb (\$60/ton, dry basis); X_{cell} = 0.4 (cellulose fraction in substrate roasting C_E); Y_{GE} = 0.8 × 1.11 = 0.89 (80% conversion in cellulose to glucose corrected for 11% weight gain due to water added by hydrolysis); C_p = \$0.005/lb (hypothetical pretreatment cost); W_{cat} = 200 enzyme units (U)/liter (hypothetical enzyme level); C_E = \$20 per 10^6 U (hypothetical enzyme cost); and G_E = 89 g glucose/liter ÷ 454 = 0.196 lb glucose/liter (final sugar concentration). Thus, the major costs, added together, are

$$T_{cost} = \frac{(\$0.03 + \$0.005)/lb}{(0.4)(0.89)} + \frac{(200\ U/liter)(\$20 \times 10^{-6}/U)}{0.196\ lb/liter}$$

$$= 0.0983 + 0.0204 = \$0.1187/lb$$

In terms of cost, an enzyme level of 200 U/liter (at \$20/10^6 U) is equivalent to 70 g/liter H_2SO_4 (as catalyst) at \$0.026/lb (unit cost including acid neutralization after hydrolysis).

This particular example shows the major cost is the biomass cost. Consequently, as previously stated, coproduct credits for pentoses (from hemicellulose hydrolysis) and lignin will be essential for reducing glucose costs. Calculations for various cases are summarized in graphical form elsewhere (Ladisch et al., 1983).

8. COMPARISON TO GLUCOSE FROM STARCH

The production of glucose from corn involves fractionation of starch (about 70% of corn dry weight) from the protein, oil, and fiber fractions, followed by starch hydrolysis. At \$3.20/bushel corn (\$134/dry ton), the glucose cost is \$0.088/lb and enzyme cost is \$0.0042/lb for a total of \$0.092 (Ladisch et al., 1983; Borglum, 1980). When by-product credits (equivalent to \$0.032/lb) are subtracted, the net cost is \$0.06/lb glucose. Additional credits may be realized by applying cellulose hydrolysis technology to the fiber fraction (Voloch et al., 1984). This comparison gives a goal that must be attained for glucose from biomass to be competitive to glucose from starch.

REFERENCES

Borglum, G. B. In Klaas, D. G. and Emert, G. (eds.), *Fuels from Biomass and Wastes,* MI, Ann Arbor Science, 1980, pp. 297–310.

Bungay, H. R. In Von Weizsacker, E. V., Swaminathan, M. S., and Lemma A. (eds.), *New Frontiers in Technology Applications: Integration of Emerging and Traditional Technologies,* Tycooly Int., Dublin, 1983, pp. 89–93.

Chase, J. *J. Polymer Sci.* **C36,** 343, 1971.

Cowling, E. B., and Kirk, T. K. *Biotech. Bioengr. Symp.* **6,** 95, 1976.

Fengel, D. *J. Polymer Sci.* **C36,** 383 1971.

Ladisch, M. R., Lin, K. W., Voloch, M., and Tsao, G. T. Process considerations in the enzymatic hydrolysis of biomass, *Enzyme Microb Technol* **5**(2), 82–102, 1983.

Reese, E. T., Lin, R. G., and Levenson, H. S. *J. Bacteriol* **59,** 485, 1950.

Tsao, G. T., Ladisch, M. R., Voloch, M., and Bienkowski, P. Production of ethanol and chemicals from cellulosic materials, *Process Biochem.* **17**(5), 34–38, 1982.

Voloch, M., Ladisch, M., Bienkowski, P., and Tsao, G. T. Bioutilization of cereal lignocellulose, in Rasper, V. F. (ed.), *Cereal Polysaccharides in Nutrition and Technology,* St. Paul, MN, Am. Assoc. Cer. Chemists, 1984, pp. 103–125.

Wenzl, H. F. J. *The Chemical Technology of Wood,* New York, Academic Press, 1970.

Whistler, R. L., and Richards, E. L. Hemicelluloses, in Pigman and Horton (eds), *The Carbohydrates,* vol. IIA, New York, Academic Press, 1970, pp. 447–467.

5

APPLIED GENETICS FOR BIOCHEMICAL ENGINEERING: RECOMBINANT DNA

MICHAEL H. HANNA

Associate Professor of Biology
Rensselaer Polytechnic Institute
Troy, New York

Over the past several years advances in the techniques and the applications of recombinant deoxyribonucleic acid (DNA) have caused a revolution in applied biology. Beginning with an understanding of restriction and modification of DNA that helps explain variable bacteriophage growth in different host strains of *Escherichia coli* and continuing with the isolation of the first type II restriction endonuclease in 1970 (Smith and Wilcox, 1970), a whole new field of genetic engineering was established. The discovery that any two pieces of DNA could be joined (Mertz and Davis, 1972) to give new gene combinations (chimeras) not found in nature has led to the use of recombinant DNA in both basic and applied research. In only a few years we have seen the development of engineered insulin (Chau et al., 1979; Goeddel et al., 1979). In 1982 recombinant human insulin was approved for use in four countries, including the United States. Other headline-catching developments include the cloning of various interferons (Hitzeman et al., 1983), development of an engineered vaccine for foot and mouth disease (Kupper et al., 1981), and cloning of the hormone somatostatin (Itakura et al., 1977); these are only a few relevant industrial applications of the new genetics. The potential uses of recombinant DNA are still vastly untapped, with the chemical, pharmaceutical, food, and biochemical industries just beginning to develop new products or techniques using recombinant DNA.

The area of applied genetics is multidisciplinary, with microbiologists, geneticists, molecular biologists, and biochemical engineers all involved in basic research and product development. Modern biochemical engineers should be familiar with the techniques involved in recombinant DNA research.

This chapter provides an understanding of the basic techniques of recombinant DNA research; it cannot, however, be totally comprehensive in showing all permutations of the techniques. Since obtaining a product is an important industrial consideration, another part of this chapter will discuss strategies and problems involved in getting recombinant DNAs to be expressed (or make a product). Finally, the remainder of the chapter will explore a relatively new area for research in which specific enzymes may be engineered for use in the biotechnology industry.

1. THE BASICS OF RECOMBINANT DNA TECHNOLOGY

Areas that will be discussed include the methods for splicing DNAs, choice of vectors and hosts, preparation of DNA for cloning, and selection of the transformed host containing the cloned DNA. Methods used to improve expression of the cloned DNAs will be discussed separately.

1.1. Methods for Gene Splicing

There are several detailed methods for splicing two pieces of DNA together, but almost all of the methods are based on three general techniques: sticky-end splicing using restriction endonucleases, artificial production of complementary ends, and blunt-end ligation (see Watson et al., 1983).

1.1. Sticky-End Splicing Using Restriction Endonucleases

A large number of type II restriction endonucleases is commercially available. Many of these restriction endonucleases cut both strands of the DNA within a recognition sequence having a twofold axis of symmetry. The particular enzymes used for sticky-end splicing make staggered restriction endonuclease cuts that yield duplex DNAs with single-strand tails containing complementary base sequences at each end of the cut. These complementary ends allow any two DNA molecules that have been treated with the same restriction endonuclease to be joined (Mertz and Davis, 1972) because the four to six base complementary regions can hydrogen bond to other

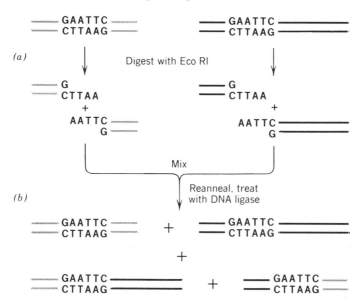

Figure 5.1. Splicing two different DNAs treated with a restriction endonuclease that leaves complementary ends. Two different DNA molecules are shown containing the six base pair recognition sequence for the restriction endonuclease EcoRI. After treatment with the restriction endonuclease, the DNA molecules are cleaved within the symmetrical recognition sequence, leaving a four base complementary DNA sequence (or single-stranded tail) at each end of the molecule. If all of these EcoRI-treated molecules are mixed, allowed to reanneal, and treated with DNA ligase to reseal the sugar phosphate backbone of the DNA molecule, four DNA combinations are possible. The top two molecules are the original parental DNA molecules that have reassociated. The lower two molecules are new combinations of DNA, or chimeric molecules, where one portion of each parental DNA molecule is associated with a portion of the other parental DNA molecule.

similarly treated DNAs (Fig. 5.1). If the hydogen-bonded DNA molecules are then treated with DNA ligase (generally T4 DNA ligase), the two DNAs will be covalently linked into one molecule. There are over 200 known restriction endonucleases that recognize approximately 60 different four to six base pair (bp) sequences. A variety of different regions or sequences in the DNA can be obtained simply by varying the enzyme used to cleave the two DNAs that are to be spliced.

There are certain disadvantages in using this method of gene splicing. The first involves the vector that is used to carry the spliced DNA. Both the vector and the foreign DNA must be cut using the same restriction endonuclease. If the chosen restriction endonuclease makes multiple cuts in the vector, many smaller or permuted arrangements of the vector DNA and the foreign DNA will be obtained. For optimum efficiency the restriction endonuclease should cut the vector only once. This ensures that a complete vector is reformed and places the recombinant DNA at a known location. Thus the vector limits the range of restriction endonucleases that can be used to obtain the recombinant piece from the foreign DNA. A second disadvantage is that the restriction endonuclease recognition site may occur within the gene of interest, allowing only a portion of the gene to be spliced into the vector. This can be overcome using partial restriction endonuclease digests, but it is a complication. Generally, restriction endonucleases having longer recognition sequences will cut a DNA molecule less frequently and be of more use. There are also problems in that either the foreign DNA or the vector may reanneal with themselves and not with each other. This problem is partially avoided by treating the vector with bacterial alkaline phosphatase to remove a phosphate residue from the 5′ end of the cleaved vector. DNA ligase will not reseal the vector, but foreign DNA will covalently link to the vector by the action of DNA ligase. Each strand of the chimeric DNA will have a single cut in the DNA backbone that is not repaired by the action of DNA ligase. However, this DNA will circularize and remain intact because of the length of the overlap region. It is effective in transformation of the host strain even though it has a nick in each DNA strand. The major advantages in using the sticky-end technique for gene splicing are that cloned DNAs can be recovered by treating the recombinant with the same restriction endonuclease used during its construction, and there is high efficiency of formation of chimeric DNAs.

1.1.2. Artificial Production of Complementary Ends

The DNA to be cloned may be obtained by random shearing of the DNA, as a complementary DNA (cDNA), or as a synthetic gene (explained in 1.3.3. and 1.3.4.). In these cases, there are no complementary ends that can be combined with a restriction endonuclease cut vector. However, complementary ends can be built onto these DNAs (Lobban and Kaiser, 1973). The enzyme terminal deoxynucleotidyl transferase can be used to add a homopolymeric tail onto the 3′ OH end of the DNA. This enzyme requires a small single-stranded 3′ end and does not require a template to add nucleotides. If the vector is treated with a restriction endonuclease that cuts only once, a homopolymeric tail can be added onto it (PstI works well because it leaves 3′ single-stranded tails). For example, if the vector has deoxyguanosine residues and the foreign DNA has deoxycytosine residues (deoxyadenosine with deoxythymidine also works) added enzymatically, the tails will be complementary and will anneal by hydrogen bonding (Fig. 5.2). Often the alignment is not perfect, and it is necessary to treat these chimeric DNAs with DNA polymerase I to fill the gaps before ligating the pieces together. This method ensures that the vector cannot interact with itself (nor can the foreign DNA) and that the vector will only interact with the foreign DNA having the complementary tail. Although DNAs obtained in any manner can be spliced using this method, one major disadvantage is that it is difficult, and in some cases impossible, to get the spliced DNA back out of the vector. It is possible, using certain restriction endonucleases and the proper homopolymeric tailing, to be able to recover the cloned DNA from the vector. For

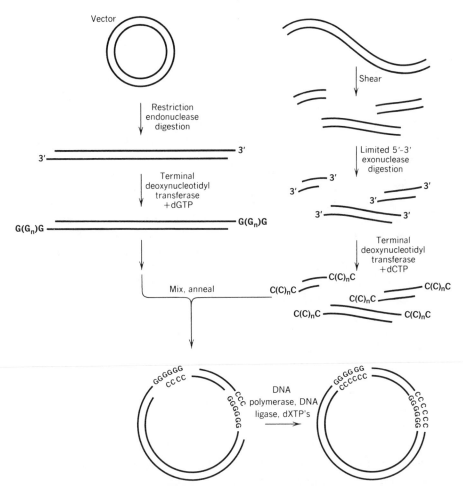

Figure 5.2. DNA splicing using complementary homopolymeric tails. The vector is treated with a restriction endonuclease cleaving it once to leave a 3′ OH single-stranded end. Terminal deoxynucleotidyl transferase plus dGTP is added for synthesis of polydG sequences at the 3′ ends. The DNA to be spliced into the vector is sheared and treated with a 5′–3′ exonuclease to provide short 3′ single-stranded ends. Terminal deoxynucleotidyl transferase plus dCTP is used to add polydC residues at the 3′ ends of the sheared DNA. If the vector and the sheared DNAs are mixed and allowed to anneal, they will only associate with each other and not self-associate. Any gaps in the chimeric DNAs are filled using DNA polymerase, and the DNAs are joined using ligase.

example, cutting the vector with PstI and tailing it with deoxyguanosine will regenerate the PstI recognition site. Any added DNA (has deoxycytosine tail) can be recovered from the chimeric vector by treating it with PstI. Another disadvantage of this technique is that additional base pairs are placed between the vector and the foreign DNA. The additional DNA may affect expression of the cloned DNA.

1.1.3. Blunt-End Ligation

Certain type II restriction endonucleases cut the DNA at the axis of symmetry within the recognition sequence (for example, AluI), thus generating blunt ends. These blunt-ended pieces of DNA (no single-stranded regions at the ends) can be spliced using T_4 DNA ligase. This process is relatively inefficient, requiring high concentrations of the ligase and the DNA to be blunt-end ligated. Blunt-end ligation can be used with restriction endonucleases that yield single-stranded ends if the single-stranded ends are filled in using DNA polymerase I or removed with S1 exonuclease before ligation. Filling in the tails regenerates duplex DNA having the restriction endonuclease site and allows splicing of DNAs that have been obtained with two different restriction endonucleases. Many times blunt-end ligation is used to add synthetic oligonucleotides to DNA (Fig. 5.3). Generally, these synthetic oligonucleotides contain recognition sites for restriction endonucleases. For example, a six base pair EcoRI restriction endonuclease recognition sequence is blunt-end ligated to a sheared DNA fragment. This fragment can be spliced into a vector by treating the blunt-end DNA containing the synthetic oligonucleotide and the vector with EcoRI. Both DNAs have complementary sticky ends and can be joined together. The only disadvantages to blunt-end ligation are its inefficiency and the fact that since any two pieces of DNA can be joined together, it is necessary to characterize the recombinant in some way to determine which fragments have been joined together.

1.2. Hosts and Vectors

1.2.1. Prokaryotic Hosts

The most frequently used bacterial hosts are *E. coli, Bacillus subtilis, Pseudomonas* species, and *Streptomyces. Bacillus subtilis* has the advantage of not being pathogenic for humans and of being good for excretion of proteins. Both features are useful in developing cloning systems. *Escherichia coli* is the most frequently used host for recombinant DNA work because its biochemistry and molecular biology are better understood than that of any other microorganism. For *E. coli* several safe strains (will not survive in nature) are available and are classified as EK2 containment hosts (see National Institute of Health guidelines). There are several problems inherent in the use and maintenance of recombinant strains. Growth yields of many recombinant strains are low. Also, cloned DNAs are frequently unstable, leading to problems in maintaining stocks after cloning. There is a general problem when eukaryotic DNA is cloned into prokaryotes involving expression of the cloned eukaryotic genes. Prokaryotes will not process eukaryotic messenger ribonucleic acids (RNAs) and will not process or modify eukaryotic proteins when they are made. With this in mind, it should be pointed out that the human insulin genes were cloned in *E. coli*. It is possible

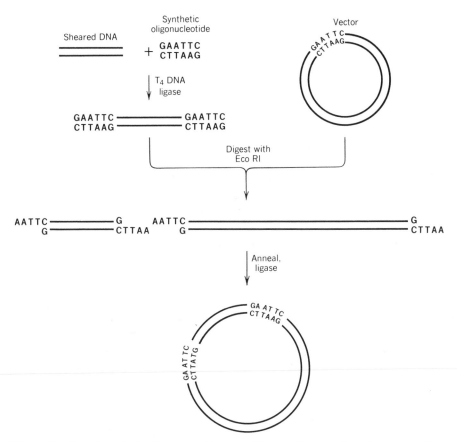

Figure 5.3. Use of synthetic oligonucleotides containing restriction endonuclease recognition sites and blunt-end ligation for splicing DNAs into a cloning vector. A six-base pair synthetic oligonucleotide containing the recognition sequence for EcoRI is blunt-end ligated onto each end of a sheared DNA using T_4 DNA ligase. This DNA and a cloning vector having a single EcoRI recognition site are mixed and digested with EcoRI (arrows indicate cleavage sites). The two pieces of DNA are spliced together based on their complementary ends.

to overcome problems of expression of eukaryotic genes in prokaryotes, but special techniques are required. These will be discussed in section 2.

The vectors used for prokaryotic cloning are either plasmids or viruses having the following general features: They generally have relaxed replication control, meaning that the plasmids are present in multiple copies per host chromosome, whereas viruses capable of lytic replication have large numbers of DNA copies present after infection. The vectors usually have a single recognition sequence for several different restriction endonucleases, providing some choice for designing cloning schemes. There is some unique way of recognizing the chimeric vector after transformation of the host.

Last, these vectors are engineered such that they are unable to transfer their DNA under normal growth conditions.

Two of the most widely used plasmids are pBR322 and pMB9. Both are derivatives of the colicinogenic plasmic colEl. Genes carried by pBR322 encode resistance to the drugs ampicillin and tetracycline to allow for selection after transformation of the *E. coli* host. It also has single recognition sequences for six different restriction endonucleases, including EcoRI, PstI (within the ampicillin resistance gene), and BamHl (within the tetracycline resistance gene). It is small, only 4.4 kb pairs, and can be amplified such that a cell containing pBR322 treated with the drug chloramphenicol will accumulate up to 3000 copies of the plasmid DNA. Amplification of the vector (host chromosome is not replicated in the presence of chloramphenicol) provides a way to obtain large amounts of recombinant DNAs. If the foreign piece of DNA is cloned into one of the drug-resistance genes, transformants containing this recombinant plasmid can easily be screened (Fig. 5.4). The transformants will be resistant to only one drug having lost their sensitivity to the other drug (where foreign DNA is located). If long pieces of DNA are inserted into plasmid vectors, the chimeric vectors are less stable in the hosts. The decreased stability may be caused by smaller plasmids having a selective replication advantage over larger plasmids.

For cloning larger pieces of foreign DNA it is often better to use the modified phage lambda vectors (Thomas et al., 1974; Leder et al., 1977) and an *in vitro* system for processing the phage DNA (Hohn and Murray, 1977) into a mature virus. Normally, lambda DNA is replicated in long concatomers after infection of the cell. In order to package one lambda DNA molecule into an empty phage head, approximately 35–45 kb pairs of DNA are required to fill the head. A specific endonuclease recognizes a DNA sequence (called the cos site) that is present every 35–45 kb pairs in the concatomer and cleaves the DNA. The cutting at the cos site leaves a sticky-ended sequence that allows the viral DNA to circularize after infection of a new host. The components necessary for packaging lambda DNA into mature phage can be obtained and used in an *in vitro* packaging system. Certain lambda strains have been engineered to have fewer restriction endonuclease

---

Figure 5.4. Identification of *E. coli* transformed with pBR322 containing inserted DNA. (*a*) pBR322 contains genes coding for resistance to both ampicillin and tetracycline. Digestion of pBR322 with BamHI cuts the DNA within the *tet* gene. Foreign DNA digested with BamHI can be spliced into the vector. There are two relevant outcomes for the vector; either it reseals, generating the original vector (Amp^r Tet^r) or a piece of DNA is inserted in the vector inactivating the *tet* gene (Amp^r Tet^s). (*b*) If these plasmids are mixed with *E. coli* and Ca^{2+}, a low percentage of the bacteria will become transformed. Any bacterium that receives either plasmid is resistant to ampicillin (Amp^r), whereas nontransformants remain sensitive. Plating the bacteria on medium containing ampicillin will allow only the transformants to grow. Transformants having a DNA insert will be sensitive to tetracycline (Tet^s) and by replica plating onto medium with tetracycline transformants having inserts can be distinguished. These transformants (indicated by open circles, Amp^r Tet^s) are recovered from the ampicillin plate.

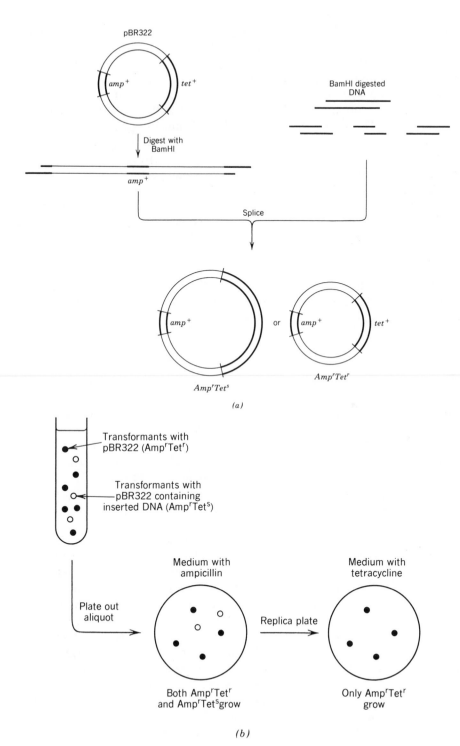

(a)

(b)

111

sites and be smaller than wild type phage. Up to a 15-kb pair DNA fragment with restriction endonuclease sites at either end can be spliced into the lambda DNA. If this DNA is used in an *in vitro* packaging system, the lambda containing recombinant DNA of the appropriate size will be packaged into the phage head. It is important that lambda-containing recombinant DNA be approximately 35–45 kb pairs in length and have cos sites in order for the DNA to be packaged properly into the phage head. These features of lambda maturation ensure that only a recombinant DNA molecule is propagated since the phage without an insert is too small to be packaged. The cos sites are also used with plasmid vectors to make cosmids (Hohn, B., 1979; Collins, J., 1979). A cosmid has the cos site of lambda cloned into a plasmid (for example, into the ampicillin-resistance gene of pBR322). If foreign DNA is cloned into this vector, larger pieces of DNA can be cloned than in a lambda alone (pBR322 plus the cos site would be less than 5 kb pairs). Using an *in vitro* system for packaging lambda DNA, a cosmid of from 35–45 kb pairs can be packaged and used to infect *E. coli* where it is able to function as a plasmid. The recombinant DNA packaged into a lambda phage head is very efficiently introduced into a host by infecting the cells as in a normal lambda infection.

1.2.2. Eukaryotic Hosts

The yeast, *Saccharomyces cerevisiae* is the most frequently used eukaryotic host for recombinant DNA studies (Hinner et al., 1979), although mammalian cells in tissue culture can be used. Yeasts have certain useful properties that make them good hosts, including the fact that eukaryotic genes will be expressed without some of the complications found in prokaryotic systems. Yeast will grow to a very high density under normal conditions and do not produce any toxins. They also secrete a number of proteins. They have endogenous plasmids (2 μm) and recombinant vectors that are very stable when transformed into yeast cells. Transformation techniques using spheroplasts are available.

Many yeast-cloning vectors contain portions of the 2-μm plasmid. This plasmid has a single replication origin, is 6.3 kb pairs, and is present in about 50 copies per cell. It has an autonomous replicating sequence (ARS) and two proteins that function to maintain the plasmid number. If the copy number of the 2-μm plasmid goes much below 50, the two plasmid proteins trigger additional replication until the copy number is returned to normal. The plasmid is stable because it has a high copy number and is not diluted out during cell division. Prokaryotic plasmids can be used to transform yeast; however, the transformants are not stable unless the recombinant DNA is integrated into a yeast chromosome. The instability of the prokaryotic plasmids results from the fact that they do not have an ARS sequence or replication origin that is recognized in yeast. The ARS sequences have been found in the 2-μm plasmid and within the yeast chromosomes. If an ARS sequence is cloned into pBR322, replication will occur but the plasmid is still not stable. A vector containing a cloned yeast centromere is likely to be stable since it will segregate during mitosis. The centromeres from three yeast chromosomes have been cloned (Fitzgerald-Hayes et al., 1982). An optimal yeast vector should be able to replicate, have an ARS sequence and cloned centromere, and carry a yeast gene that complements a deficiency in the host to simplify detection of the transformants. If a centromere region and an ARS sequence are cloned into pBR322, the plasmid will be retained stably by the yeast. By combining part of pBR322 with part of the yeast 2-μm plasmid, a vector can be made that will replicate in both *E. coli* and yeast. These vectors are called shuttle vectors (Hicks et al., 1982) and are extremely useful for cloning eukaryotic genes. Shuttle vectors have replication origins for *E. coli* and yeast and can be detected both in bacteria (drug resistance) and yeast (auxotrophic complementation). These vectors allow the special cloning features of pBR322 to be used in obtaining the spliced DNA, which then can be used to transform yeast where it will be expressed.

Transformation of mammalian cells generally involves complementation of a thymidine kinase deficient (tk⁻) host with a normal tk gene carried on a vector. The vector of choice is the transforming virus SV40 (Jackson et al., 1972). In using SV40 as a vector, either the genes expressed early or late after infection are replaced by the foreign DNA. This yields a defective recombinant SV40 that requires a helper virus for coinfection. Both the bovine papilloma viruses and retroviruses can be used as cloning vectors; however, SV40 is probably the most well characterized of all of the mammalian tumor viruses.

1.3. Preparation of DNA for Cloning

DNA to be used in cloning experiments may be obtained in one of several ways. First, fragmentation of the entire genome can be used for "shotgun" cloning experiments where it is important to get as many different pieces of a genome spliced into a vector. Second, specific genes may be selected directly from the genome and then spliced into a vector. Third, a copy of a gene may be obtained by making a copy of the messenger RNA (mRNA) for that gene using the enzyme reverse transcriptase. Fourth, a totally synthetic gene can be obtained when a small peptide is the end product of the gene.

1.3.1. Fragmentation of the Genome

The DNA of any organism is usually quite sensitive to hydrodynamic shear. Sheared DNA may be obtained by stirring the DNA or by expelling it from a syringe. It is important to consider the size of the DNA pieces. Since shearing is a random process, fragments of 15–20 kb pairs will usually be large enough to contain entire genes, except for eukaryotic DNAs where fragments up to 40 kb pairs may be necessary. Restriction endonucleases can be used to generate large DNA fragments under conditions of limited digestion. Restriction endonucleases having short (four base pair) recognition sequences will approach a random distribution for fragmentation (for example, HaeIII and MboI). These DNAs can be spliced into a plasmid or a phage by adding complementary tails using one of the methods discussed in Section 1.1.

Generation of a gene library is often useful when genes from one organism will be used repeatedly (Clarke and Carbon, 1976). It is easier to screen the library for the gene than to clone the DNA from the organism repeatedly. Sheared DNA is generally used for creating gene libraries since it provides a random population of pieces. In making a library it is important to determine that all sequences of the DNA are present. We can estimate how many transformants are necessary based on the following mathematical formula: $N = \ln(1 - P)/\ln(1 - L/G)$, where N is the number of transformants, P is the probability that every sequence is present in the library, L is the average molecular weight of the sheared fragments, and G is the molecular weight of the genome (Clarke and Carbon, 1976). For yeast DNA, if we assume a 99% probability and a fragment molecular weight of 8×10^6 (or 12 kb pairs), approximately 6000 transformants are required so that any yeast gene will be represented in the library with a probability of 99%. Almost all genes have an equal probability of being cloned with the exception of highly repetitive DNAs and regions of the genome in which a sequence in the DNA is repeated one after another (tandemly repeated DNAs). These DNAs are often less stable in the transformants since the opportunity for recombination and rearrangement is high.

1.3.2. Direct Selection from the Genome

Direct selection from the genome (Maniatis et al., 1978) is easiest with repetitive genes, but can be used for nonrepetitive DNAs to generate partial genomic libraries starting with a DNA population enriched for the sequence desired. If sheared or restriction endonuclease-treated DNA is subjected to electrophoresis on an agarose gel, a continuous smear of different sized DNA pieces is obtained from top to bottom of the gel. Using the Southern blotting technique (Southern, 1975) this DNA can be transferred to nitrocellulose paper and probed either with a radioactive mRNA for the gene in question or with an enriched mRNA population. The fragments that hybridize to the mRNA can be further purified, or the region of the gel identified by mRNA hybridization can be used to make a sub-library of the DNA in that region. The library will be enriched for the gene in question. The direct selection method is also used to isolate a genomic sequence that corresponds to a cDNA (explained in the next section).

1.3.3. Complementary DNA

Most eukaryotic cytoplasmic mRNAs have polyA tails at their 3′ ends. If a purified cytoplasmic mRNA can be obtained (or an enriched mRNA population), a DNA copy of this mRNA (or a cDNA) is obtained using the enzyme reverse transcriptase (Verma, 1977; Okiyama and Berg, 1982). Reverse transcriptase converts an RNA template into a DNA molecule. Hybridization of a polydT oligonucleotide to the polyA region of the mRNA allows reverse transcriptase to synthesize a full-length DNA copy of the mRNA. Frequently, less than full-length copies are obtained, but under certain conditions the number of full length copies is increased. These DNA:RNA hybrids can be converted into duplex DNA using either reverse transcriptase or DNA polymerase I from *E. coli* and S1 nuclease. The double-stranded DNA is ligated into a vector after complementary tails are added or by blunt-end ligation. There are certain advantages in using cDNAs when eukaryotic genes are being cloned into prokaryotic hosts. The cytoplasmic mRNA used to make the cDNA has already been processed *in vivo* and will be more easily expressed than a genomic DNA where the mRNA must be processed to remove introns before it is translated. Starting with cDNA avoids the problem that prokaryotic hosts are unable to process eukaryotic mRNAs.

1.3.4. Synthetic Genes

For small genes it is possible to synthesize the DNA (Gait and Sheppard, 1977; Smith, 1983). Both the A and B peptide genes (63 bp and 90 bp, respectively) for human insulin (Goeddel et al., 1979) and the somatostatin gene (Itakura et al., 1977) were made using chemical methods. Assuming the amino acid sequence of a peptide is known, a codon sequence in the nucleic acid can be assigned. In assigning the codons for the amino acids there is a certain amount of choice because of the degeneracy of the genetic code. In any host, certain codons are preferentially used; thus for good expression it is important to choose codons that are well used in the host in which the gene is cloned. The easiest way to synthesize the oligonucleotides in question is to use an automated system. These systems cost between $25,000 and $45,000 and can be used to make oligonucleotides up to 20 bases in length. The automated synthetic systems link the 3' OH group of the first nucleotide to a solid support, use protective groups on the added nucleotides to ensure proper bond formation between nucleotides and add bases sequentially. If the gene being synthesized is longer than 20 bases, it is made in pieces (both strands are synthesized) that are added together by blunt-end ligation. The final synthetic gene often has an initiation codon at the beginning and a termination codon at the end. A variety of different splicing methods is used to put the synthetic gene into the vector. Chemical synthesis is important in making oligonucleotides containing restriction endonuclease recognition sites (Scheller et al., 1977) for blunt-end ligation to sheared DNAs or cDNAs. The major disadvantages with chemical synthesis involve purification of each oligonucleotide after synthesis and the small size of the product.

1.4. Selection of the Transformed Host Containing the Cloned DNA

Recombinant DNAs are introduced into the host using transformation with plasmid DNAs or transfection with viral DNAs. If the recombinant DNA is packaged into a mature phage, standard infection techniques are used to introduce the virus into the host. Most procedures for transformation with prokaryotic hosts use high concentrations of Ca^{2+} and circular DNA molecules (Oishi and Irbe, 1977) because they are less susceptible to nucleases. In the eukaryote yeast (Hinner et al., 1979), removing most of the cell wall using lytic enzymes yields a spheroplast. The recombinant DNA, Ca^{2+}, and polyethylene glycol are added to get uptake of the DNA into the yeast. The yeast cells are then allowed to regenerate a cell wall. The efficiencies of transformation in prokaryotes and eukaryotes are low, and a selection technique is necessary to identify the transformed cells.

1.4.1. Detecting Plasmid Transformed Cells

Most of the prokaryotic plasmids used as vectors carry drug-resistance markers. For example, pBR322 contains a gene for resistance to ampicillin and one for resistance to tetracycline. Any cell that has been transformed with this plasmid is resistant to both drugs, and growth of the nontransformed bacteria is repressed by one or both of the antibiotics. Many cloning schemes use a restriction endonuclease that will insert the foreign DNA into one of the drug-resistance genes. This insertion causes a loss of function for that drug-resistance gene as illustrated in Figure 5.4. In eukaryotic systems, the most frequently used method involves complementation by the vector of an auxotrophic mutant host. In yeast, vectors containing either tryptophan or leucine genes have been used to complement a host that is deficient in production of the appropriate amino acid. Growth in medium not containing the amino acid will select against nontransformants. These are only two examples of screening techniques. Many others can be worked out depending on the vector and host systems used. Selection techniques may follow strategies similar to those used for selection of genetic mutations. The viral systems that were described with lambda guarantee that each recovered mature phage contains a recombinant DNA due to the maturation and packaging of the DNA. No further selection is necessary for these phage, and standard infection techniques are used to introduce the DNA into the host.

1.4.2. Determining that the Transformed Cell Contains the Gene of Interest

If a pure cDNA or restriction endonuclease fragment is inserted into pBR322, the sequential drug selection technique described above will suffice to locate the DNA. Characterization of the size of the inserted DNA in the plasmid using electrophoresis in agarose gels and hybridization of the plasmid to purified mRNA (using Southern blotting) will conclusively establish the presence of inserted DNA in the transformed host. Hybridization techniques can be used to screen genomic or cDNA libraries to identify the transformants carrying the gene of interest provided that a purified radioactive mRNA is available. The easiest method is a modification of the Southern technique (Grunstein and Hogness, 1975) in which individual colonies or phage plaques containing single recombinant DNAs are replica plated to a filter paper. The papers are treated with alkali to lyse the cells, denature, and fix the bacterial or viral DNA onto the filter. The radioactive mRNA probe is added to the paper, and only specifically hybridized mRNA is not removed by washing. Autoradiography is used to determine which colonies have DNA corresponding to the purified mRNA probe. A variation of the technique can be used to detect immunologically the colonies producing the protein coded by the recombinant gene (Broome and Gilbert, 1978). For this procedure to work, a purified mRNA is not required, but antibodies directed against the purified protein are necessary.

The techniques above will work nicely if a purified gene is cloned or if a purified mRNA probe or an antibody against a protein is available. It is much more difficult to find a transformant if these conditions are not met. However, there are a few techniques that will help. If the gene of interest has an observable phenotype, it may be possible to screen for it. For example, an *E. coli* auxotroph can be transformed, and prototrophic colonies that are drug resistant (assume library is in pBR322) are likely to contain the gene complementing the auxotrophic mutation within the vector. Complementation of a defective host is more difficult when eukaryotic genes are cloned into prokaryotic hosts (globin genes are not present in *E. coli* and could not be complemented). However, approximately 30% of the yeast genes tested are able to complement *E. coli* mutants. Complementation will work only if the cloned gene is efficiently transcribed and translated in the new host.

It is likely, based on the statement above, that many yeast genes do not contain introns. There are additional steps that can be used in developing cloning procedures for less abundantly expressed genes. The first procedure involves fractionation of total mRNA using a sizing procedure such as sucrose gradient sedimentation. The different-sized mRNA populations are added to an *in vitro* translation system, and the proteins synthesized are analyzed to determine what size class of mRNA is responsible for synthesis of the desired protein. Further size fractionation of the relevant class can be accomplished with better resolution using agarose gel electrophoresis. In this manner, more purified mRNA populations are obtained to use in cDNA cloning. In addition, gene libraries can be screened using synthetic DNA oligonucleotide probes (Suggs et al., 1981). If a 10–20 amino acid sequence for a protein is known, a mixed oligonucleotide probe (discussed in Section 3) can be synthesized and used to screen the gene library or as a primer to begin DNA synthesis with reverse transcriptase on mRNA populations. The cDNAs made using this probe should be highly enriched for the gene and are used to make a library in an appropriate vector.

2. EXPRESSION OF RECOMBINANT DNA

The ultimate goal in the application of genetic engineering is to obtain a product, in most instances a protein, that is either difficult to obtain under standard conditions or is present in very low amounts. The first problem is obtaining the gene coding for this product and splicing it into an appropriate vector. Assuming that the cloned gene is available, it is necessary to ensure that it is expressed in the new host. Expression of any gene is dependent on efficient transcription of the gene into mRNA and efficient translation of that mRNA into a protein product. There are several problems encountered when expression of the DNA is important. First, when a gene is inserted into a vector it may be in one of two orientations, only one orientation yielding the sense transcript of the gene. Second, each cloned gene must be preceded by a promoter sequence recognized by the host RNA polymerase for transcription to occur and a ribosome binding site and an initiation codon within the transcript to allow efficient translation of the mRNA. Third, as mentioned previously, eukaryotic genes often produce mRNAs that must be processed to remove introns. Prokaryotic hosts are unable to process these mRNAs. Fourth, bacteria are not likely to process any proteins produced from eukaryotic genes. Fifth, eukaryotic proteins are foreign in bacteria and may be preferentially degraded by bacterial proteases. Eukaryotic proteins produced in great amounts can affect growth of the host and replication of the vector causing loss of the transformant. These problems will now be discussed in greater detail.

2.1. Optimization of Expression

One general rule of thumb for obtaining expression of recombinant DNAs is that prokaryotic genes are more easily expressed in prokaryotic hosts and eukaryotic genes are more easily expressed in eukaryotic hosts. However, this rule is frequently circumvented. Regardless of the source of the gene or the host, a cloned gene has a 50/50 chance of being oriented in the plasmid in the direction required for proper transcription. Only one strand of the DNA is transcribed into mRNA; therefore, the orientation of the cloned gene in relation to its promoter is critical (Fig. 5.5). This orientation is not a problem if the cloned gene has its own promoter and regulatory sequences cloned into the vector along with the structural gene (assuming these regulatory regions are recognized by the host). However, it is a problem when eukaryotic genes are cloned into prokaryotes. Since there is a 50/50 chance of the DNA being oriented correctly, it may be necessary to check several transformants to find one that expresses the gene. Problems arise when none of

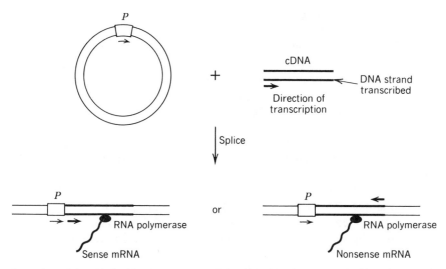

Figure 5.5. Orientation of foreign DNA in relation to a plasmid promoter. The vector containing a promoter (*P*) is shown (direction of transcription is clockwise). A cDNA is also shown. Normal transcription of the cDNA occurs on the boldface strand in the direction indicated by the arrow. Using standard splicing techniques the foreign DNA may be inserted in either direction shown in the figure. The orientation on the left has the foreign gene in the proper orientation with the plasmid promoter to obtain the "sense" transcript of the DNA, whereas that on the right gives transcription of the nonsense strand of the DNA yielding a nonfunctional mRNA.

the transformants produces a protein and it must be established why there is no expression. It is possible to use restriction endonuclease digestion of the recombinant vectors to determine the orientation of the gene in the plasmid. Depending on the orientation of the gene in the vector, different restriction endonuclease digestion patterns will be found because of asymmetry at the ends where the vector and the recombinant gene are spliced. Comparison of digestion patterns of several different transformants will allow assignment of the orientation. Certain vectors are available that guarantee proper orientation of the gene in question (Berman, 1983). The vector pUC8 can be cut with EcoRI and SalI to open up the DNA. If the recombinant gene has an EcoRI sequence added at one end and an SalI sequence at the other end, the orientation of the recombinant DNA in the vector is guaranteed.

Frequently, a cloned gene will not have its own promoter associated with it. Restriction endonuclease digestion or random shear of the DNA may separate the promoter and regulatory regions from the structural gene. cDNAs will not have promoter sequences associated, since they are copied from an mRNA. Eukaryotic regulatory sequences, if present, are not recognized by the prokaryotic transcriptional system and will cause problems with expression of the gene unless they are removed. Many *E. coli* and yeast vectors are available with promoter sequences cloned into them. The *E. coli* vectors contain strong promoters that can be regulated. Examples include the *lac* operon promoter, which can be turned off when glucose is present or on when lactose is present in the medium; fusions of the *trp-lac* regulatory sequences that are regulated by the level of tryptophan in the medium; and the lambda pL regulatory region, which can be controlled by a thermolabile repressor protein (cI857). These controllable promoters allow for expression of the genes under certain growth conditions and not under other conditions. When either induced or derepressed, they are strong promoters and transcription of the genes by RNA polymerase is extensive. Also, controllable promoters allow production of proteins that are toxic to the host cells. For example, if the gene for a potentially toxic protein were cloned into a lambda expression vector that codes for a thermolabile repressor protein (cI857), the host containing the recombinant will grow normally at low temperature. Expression of the recombinant gene is achieved after the cells have grown to high density by raising the temperature. The increased temperature inactivates the repressor allowing viral gene expression.

Once the promoter is provided, the ribosome binding sequence (Shine-Delgarno sequence) and an initiation codon must be provided for efficient translation of the mRNA by prokaryotic ribosomes. The distance between the ribosome binding site and the initiation codon is also critical for maximal expression. Often both sequences are cloned along with the promoter into the vector. It is important that the foreign gene be added to the vector in exactly the right reading frame to allow translation of the message into the correct amino acid sequence. In many cloning schemes, the initiation codon is added at the beginning of the foreign DNA to get the nucleotides that follow into the correct reading frame. Alternately, when the initiation codon is in the vector, the splicing site used to insert the foreign DNA should be in close proximity to the initiation codon. This is most easily accomplished with a gene fusion (Goedell et al., 1979; Itakura et al., 1977) between the prokaryotic gene and the foreign eukaryotic gene (Fig. 5.6). For example, an EcoRI site was introduced by mutagenesis into the *lacZ* or β-galactosidase gene of the *lac* operon in a noncritical early coding region. If the promoter and regulatory sequences of the *lac* genes and the EcoRI fragment of the *lacZ* structural gene are put into a vector, foreign DNA can be added at the EcoRI site. Provided the foreign gene is in the correct register for reading the codons, a fusion peptide is formed containing approximately 25 amino acids of the β-galactosidase protein at the NH_2 terminus fused to the eukaryotic peptide. The small fused portion of the β-galactosidase protein does not usually interfere with the function of the eukaryotic protein and often will protect it from degradation. With some fusion peptides, the prokaryotic amino acids are removed by chemical methods leaving only the eukaryotic peptide. A very useful fusion peptide could possibly be constructed. In general, peptides that are excreted from bacterial cells have a sequence of hydrophobic amino acids at their NH_2 terminus. These amino acids (called a signal sequence) are cleaved as the peptide is excreted. Fusion of a recombi-

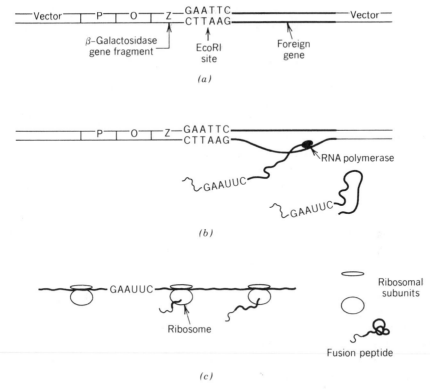

Figure 5.6. Transcription and translation of a fused gene. (*a*) Fusion gene in vector. The *lac* regulatory region (promoter operator) and a small portion of the β-galactosidase gene (*lacZ*) were added into a plasmid cloning vector. The β-galactosidase gene has a recognition sequence for EcoRI (put in by mutagenesis techniques) where foreign DNAs can be spliced into the vector, as illustrated. (*b*) Transcription of fused gene. Transcription of the β-galactosidase gene is under control of the *lac* regulatory genes, and when induced transcription begins with the *lacZ* gene and proceeds through the foreign gene. (*c*) Translation of fusion gene mRNA. Translation of the mRNA from the fusion gene is normal since the ribosome binding site and initiation codon are present in the *lacZ* gene. The fusion peptide contains only a few amino acids at the NH₂ terminus from β-galactosidase. These usually do not interfere with the activity of the foreign peptide.

nant gene to a signal sequence coding region could make purification of the protein product much easier since it may be excreted.

Almost all eukaryotic genes that have been successfully expressed in a prokaryotic host have started with a cDNA or a synthetic gene. This avoids the problem of processing a eukaryotic mRNA. Eukaryotic genomic sequences are not frequently cloned in prokaryotes when expression of the gene is desired. However, yeast has been shown to process several animal genes by removing introns from the mRNAs. Yeast may be a very important system for eukaryotic cloning in the near future. In addition, yeast RNA polymerase recognizes many animal gene promoters, making it possible to avoid specially constructed vectors containing promoter sequences. Although yeast does not process all eukaryotic proteins, it will process some of them. Another important feature of gene expression in yeast is that much less degradation of foreign eukaryotic proteins occurs than in a prokaryotic host.

2.2. Examples

The human insulin genes have been cloned in *E. coli* using the mature gene sequence and the sequence for proinsulin (for example see Goeddel et al., 1979). A synthetic gene for both the A and B peptides was made (63 bp for A and 90 bp for B). Each synthetic sequence had a methionine codon added at its beginning and a termination sequence added at the end of the gene. These synthetic genes were separately cloned into the β-galactosidase gene in an expression vector and used to transform *E. coli*. Expression of the A and B pepetide genes was controlled by the *lac* operon promoter and regulatory genes. The fusion peptides were harvested and treated with cyanogen bromide, which cleaves peptides after methionine or tryptophan residues. Cyanogen bromide treatment releases the A and B peptides. These are mixed and processed *in vitro* into mature insulin by forming disulfide bridges between the two peptide chains. Cyanogen bromide is useful for releasing the fused peptides since neither methionine nor tryptophan is present in the A or B peptides.

The rabies virus G protein is a viral surface glycoprotein that is responsible for most of the antigenic and immunogenic activities shown by the virus. Cloning and expression of this gene may be a starting point for the development of a safe vaccine for rabies. The gene coding for the G protein was obtained from the mRNA of infected cells (Yelverton et al., 1983). A cDNA library was made using pBR322 as the vector. Homopolymeric tails were added onto the cDNAs (dG) and to the vector (dC) at the PstI restriction endonuclease site. Transformants carrying the G protein gene were identified by screening with purified G protein mRNA. The DNA from the library vector was excised using restriction endonucelase digestion, and an initiation codon was added at the beginning of the gene. A 300-bp sequence containing the *trp* promoter and ribosome binding site was cloned into pBR322, and the altered G protein gene was cloned into an EcoRI site immediately adjacent to the *trp* regulatory region. The chimeric plasmids were screened for proper orientation of the gene using XbaI restriction endonuclease digestion of pBR322. The expression of the gene is under control of the *trp* operon, and a gene fusion occurred between the G protein gene and the COOH terminal coding sequence of the β-lactamase gene. The expression of the gene was improved by shortening the distance between the ribosome binding site and the initiation codon. A restriction endonuclease (XbaI) with a four base recognition sequence was used to cut the DNA between the ribosome binding site and the initiation codon. The single-stranded tails were digested with S1 endonuclease and religated with T_4 ligase. The G protein accounted for 2–5% of the cellular protein. The recombinant fusion peptide is not glycosylated (the native G protein is), although it has 70–90% of the antigenic activity of the native protein.

Several interferon (INT) genes have been cloned and expressed in a yeast cloning system (Hitzeman et al., 1983). The INT cDNAs were inserted into a yeast shuttle vector (YEp1PT). The shuttle vector contains a significant portion of the pBR322 plasmid, including the replication origin and the ampicillin-resistance gene. It also contains sequences from the yeast 2-μm plasmid necessary for stable replication and maintenance of the plasmid in yeast. The yeast *trp*-1 gene is present on the vector to allow recognition of transformants in *trp*-1 hosts. The yeast glycolytic enzyme 3-phosphoglycerate kinase (PGK) is a highly expressed gene, and its promoter and regulatory region were also added to the vector. The INT cDNAs (after adding an initiation codon) were cloned into a site next to the PGK regulatory region using EcoRI. Both mature and pre-INT cDNAs were cloned into the vector and were expressed. The pre-INT peptides have a signal sequence necessary for excretion of the peptide that is removed during excretion. Up to 30% of the pre-INT protein is excreted from yeast, and the signal sequence is not present on the excreted protein. If greater levels of excretion can be obtained, purification of the interferon from yeast will be facilitated.

3. PROTEIN ENGINEERING

Several thousand enzymes have been characterized to some extent. As many as 200 are commercially available, yet fewer than 20 are used for industrial purposes. It is likely that in the future enzymes will be used more often in industrial processes than they are now. Several recent developments support this prediction. Recombinant DNA technology, advances in protein purification, and immobilization techniques for enzymes and cells will improve the cost efficiency of enzymatic reactions as compared to similar chemical synthesis processes. These advances alone may not be sufficient to cause greater use of enzymes in industrial processes. Proteins require physiological conditions, are less stable, and catalyze very specific reactions. Most chemical processes function with fewer complications than the enzymatic processes. However, if enzymes from organisms living under a variety of environmental conditions are characterized, it becomes apparent that dissimilar proteins catalyze similar enzymatic reactions. This illustrates their range of structural diversity. If this diversity exists in nature, can it be recreated *in vitro* using recombinant DNA techniques to alter a known enzyme into a more useful industrial catalyst? To test this approach structurally, characterized enzymes should be used. Approximately 20 enzymes have been characterized by x-ray diffraction, and many others have partial structural and sequence data available. If the relationship between a structural feature and a useful function is known, it may be possible to adapt this feature to other enzymes or adapt this enzyme to a different catalytic capability.

3.1. Finding the Gene for the Enzyme

The techniques for gene cloning and expression have been described in the first two sections of this chapter. There are two techniques that are particularly helpful when the enzyme is available. First, it may be possible to locate the gene from a library made in an expression vector. With this technique individual bacterial colonies containing cloned DNAs are probed with antibodies prepared against the original protein using a modified Western blot procedure. Second, it is possible using peptide sequencing techniques to obtain a partial peptide sequence from as little as 1 μg of protein. By choosing a certain short amino acid sequence from the peptide, a nucleotide sequence of 12–20 bases can be synthesized and used to probe a gene library (by the modified Southern blot analysis described in section 1.4.2.). Since several amino acids have multiple codons, the nucleotide probe used should come from a region of the protein having few of these amino acids. For example, histidine and methionine have one codon each, whereas tryptophan, lysine, glutamic acid, and several other amino acids have two. Regions of a peptide containing these amino acids can be translated into a nucleotide sequence more easily (and with greater probability of containing the true gene sequence). These partial gene sequences are used as probes to identify the gene in a library. This technique was used to identify the rat insulin gene. Once the gene has been isolated, it should be sequenced using either the Maxam-Gilbert or the Sanger technique and cloned into an expression vector. The sequence of the gene is important for relating structural and functional information to a region of the gene that can be altered in vitro.

3.2. Altering and Designing Enzymes

The conventional method for altering the properties of an enzyme is by mutation and selection. Although these techniques have proven extremely useful, mutation is a random process that is followed by a detailed selection procedure to identify the mutant. With the use of recombinant DNA, it is possible to cause very specific mutational changes *in vitro* and alter the properties of an enzyme in a more specific manner. Using *in vitro* mutagenesis (Shortle et al., 1981) deletions, insertions, and base pair substitution mutations can be made specifically within the cloned gene. The techniques used for *in vitro* mutagenesis are simple. Deletion mutations are made at restriction endonuclease recognition sequences within the gene. The DNA is cut with the restriction endonuclease followed by digestion of the single-stranded tails and blunt-end ligation. This is a site-directed mutation. Insertions are made by treating the recombinant DNA with pancreatic deoxyribonuclease (DNase) in the presence of Mn^{2+} causing a double-strand break in the DNA (limited digestion to yield one break per DNA molecule). Synthetic oligonucleotides are added, usually restriction endonuclease recognition sequences, by blunt-end ligation to splice the oligonucleotide into the gene. Pancreatic DNase can be used for deletion mutations in conjunction with Ba131 exonuclease that degrades both DNA strands from one end of a linear DNA. Random substitution mutations are made in the DNA using a variety of chemical and repair replication techniques. Most of these techniques yield random mutations. However, the advantage over standard mutagenesis is that *in vitro* it occurs within the gene in question not throughout the genome.

Specific site-directed mutations are made using the following technique (Dolhadie-McFarland et al., 1982). A 12-base synthetic oligonucleotide complementary to the DNA sequence of the gene is made. One base in the middle of the oligonucleotide is not complementary to the gene sequence. This oligonucleotide is hybridized to a full complementary strand from the native gene under nonstringent conditions, allowing a stable intermediate to be formed. If a gene is cloned into the replicative form of the single-stranded DNA phage M13, only one strand is present in the mature phage, and this is the wild-type template strand. The mutant synthetic oligonucleotide hybridized to this strand will serve as a primer for DNA replication allowing synthesis of a complete complementary strand. This double-stranded DNA is used to transfect *E. coli*. Because of the single base pair mismatch in the DNA, either the wild-type sequence is repaired yielding a fully mutant duplex DNA or the mutant sequence is repaired yielding a fully wild-type sequence. Using the modified Southern blot technique with the original mutant oligonucleotide primer and stringent conditions for hybridization, the mutant gene can be located. The mutant gene is removed from the M13 vector and spliced into an expression vector to determine the effect of the mutation on the peptide.

How can directed mutagenesis be used to alter the properties of an enzyme specifically? It has been shown that enzymes from thermophilic microorganisms contain salt bridges, different electrostatic interactions, and modified amino acids that allow the enzymes to remain stable at higher temperatures than comparable enzymes from mesophilic organisms. The ability to function at higher temperatures depends on interactions at several sites in the enzyme. Using directed mutagenesis, these regions can be probed by mutation to determine how they affect the thermophilic properties of the enzyme. With a better understanding of what features are necessary to affect these properties, it may be possible to alter the gene for a mesophilic enzyme in a region not coding for the catalytic site in such a manner that the altered enzyme becomes more thermotolerant. More basic research on thermostability is necessary before these alterations can be fully understood. However, it is at least possible to develop alteration schemes for determining what properties are important. It is less far reaching to suggest that certain base changes in a gene can release an enzyme from allosteric regulation caused by feedback inhibition. Such mutations have been found using standard mutagenesis and may be more easily found using *in vitro* mutagenesis on cloned genes. A release from regulatory controls would allow an increased yield of product. There are also examples of mutants that have altered substrate specificities. Using *in vitro* mutagenesis, it may be easier to find these mutants, especially if the sequence and location of the active site coding regions in the gene are known. The appropriate "Buck Rogers" scenario for enzyme engineering is to have structure–function units for an enzyme and the corresponding synthetic gene sequences for these units. Then artificial enzymes having a set series of functions can be made by joining the appropriate functional units together. There is speculation that the intron–exon arrangement of eukaryotic genes separates them into coding units for different functional domains of a peptide. Nature has already learned how to splice its functional domains together; we are still lagging behind *in vitro* but not so far that we cannot see the appropriate direction to proceed.

GENERAL REFERENCES

Adams, R. L. P., Burdon, R. H., Campbell, A. M., Leader, D. P., and Smellie, R. M. S. *The Biochemistry of the Nucleic Acids,* New York, Chapman and Hall, 1981.

Birge, E. A. *Bacterial and Bacteriophage Genetics: An Introduction,* New York, Spinger-Verlag, 1981.

Freifelder, D. *Molecular Biology: A Comprehensive Introduction to Prokaryotes and Eukaryotes,* Boston, Science Books Int., 1983.

Maniatis, T., Fritsch, E. F., and Sambrook, J. *Molecular Cloning: A Laboratory Manual,* Cold Spring Harbor, NY, Cold Spring Harbor Laboratory, 1982.

Biotechnology, *Science.* **219,** 1983.

Watson, J. D., Tooze, J., and Kurtz, D. T. *Recombinant DNA: A Short Course,* New York, W. H. Freeman, 1983.

Specific References

Alwine, J. C., Kemp, D. J., and Stark, G. R. Method for detection of specific RNAs in agarose gels by transfer to diazobenzyloxymethyl-paper and hybridization with DNA probes. *Proc. Natl. Acad. Sci. U.S.A.* **74,** 5350–5354, 1977.

Berman, M. L. Vectors for constructing hybrid genes. *Biotechniques* **1,** 178–183, 1983.

Broome, S., and Gilbert, W. Immunological screening method to detect specific translation products. *Proc. Natl. Acad. Sci. U.S.A.* **75,** 2746–2749, 1978.

Chau, S. J., Noyes, B. E., Agarwal, K. L., and Steiner, D. F. Construction and selection of recombinant plasmids containing full-length complementary DNAs corresponding to rat insulins I and II. *Proc. Natl. Acad. Sci. U.S.A.* **76,** 5036–5040, 1979.

Clarke, L., and Carbon, J. A colony bank containing synthetic ColEl hybrid plasmids representative of the entire *E. coli* genome. *Cell* **9,** 91–99, 1976.

Collins, J. *Escherichia coli* plasmids packageable in vitro in λ bacteriophage particles. In Wu, R. (ed.), *Methods in Enzymology,* New York, Academic Press, vol. 68, 1979, pp. 309–326.

DeBoer, H. A., Comstock, L. J., and Vasser, M. The *tac* promoter: A functional hybrid derived from the *trp* and *lac* promoters, *Proc. Natl. Acad. Sci. U.S.A.* **80**, 21–25, 1983.

Dolhadie-McFarland, G., Cohen, L. W., Riggs, A. D., Morin, C., Itakura, K., and Richards, J. H. Oligonucleotide-directed mutagenesis as a general and powerful method for studies of protein function. *Proc. Natl. Acad. Sci. U.S.A.* **79**, 6409–6413, 1982.

Fitzgerald-Hayes, M., Buhler, J.-M., Cooper, T. G., and Carbon, J. Isolation and subcloning analysis of functional centromere DNA (CENII) from yeast chromosome XI. *Mol. Cell. Biol.* **2**, 82–87, 1982.

Gait, M. J., and Sheppard, R. C. Rapid synthesis of oligodeoxyribonucleotides: A new solid phase method. *Nuc. Acids Res.* **4**, 1135–1158, 1977.

Goeddel, D. V., Kleid, D. G., Bolivard, F., Heynecker, H., Yansura, D., Crea, R., Hirose, T., Kraszewski, A., Itakura, K., and Riggs, A. Expression in *Escherichia coli* of chemically synthesized gene for human insulin. *Proc. Natl. Acad. Sci. U.S.A.* **76**, 106–110, 1979.

Grunstein, M., and Hogness, D. S. Colony hybridization: A method for the isolation of cloned DNAs that contain a specific gene. *Proc. Natl. Acad. Sci. U.S.A.* **72**, 3961–3965, 1975.

Hicks, J. B., Strathern, J. N., Klar, A. J. S., and Dellaporta, S. L. Cloning by complementation in yeast: The mating type genes. In Setlow, J. K. and Hollaender, A. (eds.), *Genetic Engineering: Principles and Methods,* New York, Plenum, vol. 4, 1982, pp. 219–248.

Hinner, A., Hicks, J. B., Ilgen, C. and Fink, G. R. Yeast transformation: A new approach for the cloning of eukaryotic genes. In Sebek, O. K. and Laskin, A. I. (eds.), *Genetics of Industrial Microorganisms,* Washington, D.C., American Society for Microbiology, 1979, pp. 36–43.

Hitzeman, R. A., Leung, D. W., Perry, L. J., Kokr, W. J., Levine, H. L., and Goeddel, D. V. Secretion of human interferons by yeast. *Science* **219**, 620–625, 1983.

Hohn, B. In vitro packaging of λ and Cosmid DNA. In Wu, R. (ed.), *Methods in Enzymology,* New York, Academic Press, vol. 68, 1979, pp. 299–309.

Hohn, B., and Murray, K. Packaging recombinant DNA molecules into bacteriophage particles in vitro. *Proc. Natl. Acad. Sci. U.S.A.* **74**, 3259–3263, 1977.

Itakura, K., Hirose, T., Crea, R., Riggs, A. D., Heynecker, H. L., Bolivar, F., and Boyer, H. W. Expression in *E. coli* of a chemically synthesized gene for the hormone somatostatin. *Science* **198**, 1056–1062, 1977.

Jackson, D., Symons, R., and Berg, P. Biochemical method for inserting new genetic information into DNA of simian virus 40: Circular SV40 DNA molecules containing lambda phage genes and the galactose operon of *Escherichia coli. Proc. Natl. Acad. Sci. U.S.A.* **69**, 2904–2909, 1972.

Kupper, H., Keller, W., Kurz, C., Forss, S., Schaller, R., Franze, R., Strohmaier, K., Marquardt, O., Zaslavsky, V. G., and Hofschneider, P. H. Cloning of a cDNA of major antigen of foot and mouth disease virus and expression in *E. coli. Nature* **289**, 555–559, 1981.

Leder, P., Tiemeier, D., and Enquist, L. EK2 derivatives of bacteriophage lambda useful in the cloning of DNA from higher organisms: The gt WES system. *Science* **196**, 175–177, 1977.

Lobban, P., and Kaiser, D. Enzymatic end-to-end joining of DNA molecules. *J. Mol. Biol.* **78**, 453–471, 1973.

Maniatis, T., Hardison, R. C., Lacy, E., Lauer, J., O'Connell, C., Quon, D., Sim, D. K., and Efstratiadis, A. The isolation of structural genes from libraries of eucaryotic DNA. *Cell* **15**, 687–701, 1978.

Maxam, A. M., and Gilbert, W. A new method of sequenceing DNA. *Proc. Natl. Acad. Sci. U.S.A.* **74**, 560–564, 1977.

Mertz, J., and Davis, R. Cleavage of DNA: RI restriction enzyme generates cohesive ends. *Proc. Natl. Acad. Sci. U.S.A.* **69**, 3370–3374, 1972.

Oishi, M., and Irbe, R. M. Circular chromosomes and genetic transformation in *E. coli,* in

Portoles, A., Lopez, R., and Espinosa, M. (eds.), *Modern Trends in Bacterial Transformation and Transfection,* Amsterdam, North Holland, 1977, pp. 121–134.

Okayama, H., and Berg, P. High-efficiency cloning of full length cDNA. *Mol. Cell. Biol.* **2,** 161–170, 1982.

Sanger, F., and Coulson, A. R. A rapid method for determining sequences in DNA by primed synthesis with DNA polymerase. *J. Mol. Biol.* **94,** 444–448, 1975.

Scheller, R. H., Dickerson, R. E., Boyer, H. W., Riggs, A. D., and Itakura, K. Chemical synthesis of restriction enzyme recognition sites useful for cloning. *Science* **196,** 177–180, 1977.

Shortle, D., DiMaio, D., and Nathans, D. Directed mutagenesis. *Ann. Rev. Genetics* **15,** 265–294, 1981.

Smith, H. O., and Wilcox, K. W. A restriction enzyme from *Hemophilus influenzae,* I. Purification and general properties. *J. Mol. Biol.* **51,** 379–391, 1970.

Smith, J. A. 1983. Automated solid-phase oligodeoxyribonucleotide synthesis. *Am. Biotechnology Laboratory,* 15–24, 1983.

Southern, E. M. Detection of specific sequences among DNA fragments separated by gel electrophoresis. *J. Mol. Biol.* **98,** 503–517, 1975.

Suggs, S. V., Wallace, R. B., Hirose, T., Kawashima, E. H., and Itakura, K. Use of synthetic oligonucleotides as hybridization probes: Isolation of cloned cDNA sequences for human β_2-microglobulin. *Proc. Natl. Acad. Sci. U.S.A.* **78,** 6613–6617, 1981.

Thomas, M., Cameron, J. R., and Davis, R. W. Viable molecular hybrids of bacteriophage lambda and eukaryotic DNA. *Proc. Natl. Acad. Sci. U.S.A.* **71,** 4579–4583, 1974.

Ulmer, K. M. Protein engineering. *Science* **219,** 666–671, 1983.

Verma, I. M. The reverse transcriptase. *Biochim. Biophys. Acta.* **473,** 1–38, 1977.

Yelverton, E., Norton, S., Obijeski, J. F., and Goeddel, D. V. Rabies virus glycoprotein analogs: Biosynthesis in *Escherichia coli. Science* **219,** 620–625, 1983.

6

MOLECULAR ENZYME ENGINEERING

DAVID S. HOLMES

Staff Molecular Biologist
General Electric Company
Research and Development Center
Schenectady, New York

1. INTRODUCTION

Production and isolation of enzymes and the manipulation of enzymes, including enzyme immobilization, form an important and integral part of the discipline of biochemical engineering. This, of course, has been true for some years. However, despite the fact that world enzyme production in 1983 was a multibillion dollar business, the market was dominated by only a few enzyme types, notably carbohydrases and proteases (see Figure 6.1). Of the more than 2000 different enzymes that have been described, only about 200 are commercially available and only 16 or so are used in industrial amounts. Not only is there a modest number of enzymes being used for a limited number of applications, but also the market is dominated by only a handful of companies.

There are several reasons for the situation just described. Most enzymes are too expensive and many are considered too delicate for commercial use. It is frequently thought that enzymes do not exist that will carry out reactions involving water-insoluble substrates and that enzymes cannot operate on "unnatural" compounds. Also there is a degree of unfamiliarity on the

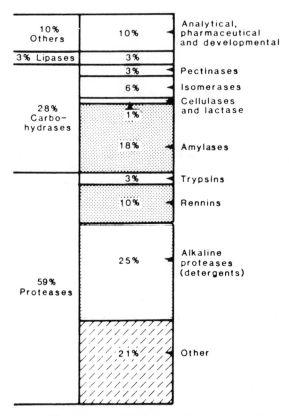

Figure 6.1. *Enzyme uses.* The major use of proteases is in detergents; other proteases are used to tenderize meat and in the production of pharmaceuticals; rennins are used in making cheeses; amylases are used to hydrolyze starch; other carbohydrases are used to produce invert sugar, to convert glucose to fructose, to hydrolyze pectic substances, and to oxidize glucose to gluconic acid. Lipases are used to hydrolyze fats and fatty esters. (After Maugh, 1984. Reprinted with permission.)

part of chemists concerning the manipulation and properties of enzymes. For example, enzymes are often considered too specific when a general catalyst is required. This is not true of all enzymes, some of which will act on quite a broad range of substrates. Last, chemical and manufacturing industries have already invested in plants that use conventional chemical catalysts, so there is an inertia to overcome in order to implement new enzyme-based processes.

However, there is a growing need for the production of complex biological and other intricate organic molecules. In these instances the potential advantages of using enzymes is becoming quite apparent. These advantages incude their:

1. Ability to catalyze the synthesis, degradation, and modificaton of chemical compounds under mild conditions

2. Stereospecificity

3. Ability to enhance complex sequential chemical reactions

4. Ability to utilize relatively inexpensive substrates

5. Production of nontoxic waste-products

6. The vast number of different enzymes available in nature

In addition, an increasing number of robust enzymes and enzymes that catalyze novel reactions have been found, including some enzymes that have catalytic activity in nonaqueous solvents. This is attracting the interest of industrial chemists.

The recognition of these advantages is coming at a crucial time when developments in other areas promise to permit the wider use of enzymes for industrial applications. These developments include advances in the design and construction of enzyme and cell reactors, advances in cell and enzyme immobilization techniques, the increasing ability to regenerate or recycle cofactors, and the spectacular advances in genetic engineering. Several of these topics are considered elsewhere in this book.

This chapter will describe some of the technical aspects of genetic engineering that have resulted in the rapid accumulation of information on enzyme structure and function and, in some instances, have even permitted the restructuring of enzymes with novel properties. This chapter will also discuss four major advances in other areas of chemistry and biochemistry that are relevant to our understanding of enzyme structure and function. It is expected that these areas will also have a major impact on the enzyme industry: (1) the use of direct, chemical modification of enzymes to alter their specificity; (2) significant progress in the elucidation of enzyme structure/function relationships through the construction of small artificial enzymes; (3) the synthesis of several nonpeptide organic molecules, termed enzyme-mimics. They exhibit many of the properties of enzymes, such as substrate specificity, catalytic activity, and reaction type. In addition they have the advantages of simplicity, robustness, and activity in organic solvents; (4) a number of naturally occurring enzymes are being shown to have a wide range of substrate specificities and reactions in addition to the biological reactions they catalyze in vivo. This is especially true if the reaction conditions are modified. Therefore, if some novel catalytic activity is desired, it may not always be necessary to look for a new or modified enzyme because presently available enzymes might be suitable under different conditions.

Before the discussion of these topics, a brief overview of the structure and function of enzymes will be given. This will serve to place in correct perspective the recent advances alluded to above. In the final section of the chapter some of the future prospects that these several advances may permit will be mentioned. Also several recent improvements in analytical techniques and instrumentation that are impinging upon enzyme engineering will be identified.

2. ENZYME STRUCTURE AND FUNCTION

An enzyme is an effective catalyst because it lowers the activation energy of one or more unstable reaction intermediates. It characteristically achieves this by virtue of its specific three-dimensional structure. It is also this three-dimensional structure that shapes the active site to fit the substrate precisely and hence achieve the remarkable specificity of an enzyme. The active site is usually a groove or concave surface containing amino acids whose side chains participate in catalysis. The concave nature of the active site permits the substrate to be sequestered into a hydrophobic environment thereby permitting strong charge interactions that would be impossible in an aqueous environment. During catalysis the substrate is bound so that the atoms participating in the bond to be made or broken are oriented properly with respect to the catalytic groups on the enzyme. Occasionally this may involve a distortion of the conformation of the substrate, facilitating bond breakage or sometimes a distortion of the enzyme itself to bring about the proper alignment of reactive groups. Recently it has become evident that catalysis often involves substantial molecular motion in and around the active site, and the concept of a rather static or rigid protein framework has been replaced by one of flexibility and motion.

Only a few of the 20 amino acids participate directly in catalysis. These tend to be those with polar groups that are good electron donors, such as serine, cysteine, histidine, lysine, aspartic acid, and glutamic acid. There are no examples of side chains that are good electron acceptors. Therefore, small molecules and metal ions (cofactors or coenzymes) with additional chemical properties are used in some enzyme reactions. Some cofactors such as thiamine pyrophosphate bind transiently to the enzyme during the reaction; others, called prosthetic groups, such as flavin adenine dinucleotide, are permanently bound to the enzyme.

Chymotrypsin, one of the family of serine proteases, exemplifies several of the characteristics of enzyme catalysis. Chymotrypsin consists of three polypeptides joined by sulfhydryl linkages and totaling 345 amino acids. The action of chymotrypsin is to hydrolyze proteins by adding water across peptide bonds. It accomplishes this in a series of stages, which include a preliminary binding of the substrate in a pocket in the enzyme. The binding of the substrate causes a molecular rearrangement of atoms in the catalytic site resulting in the activation of an oxygen on a crucial serine through a charge relay system (see Figure 6.2). The activated oxygen covalently bonds the substrate by transacylation to form a highly reactive transition complex. The substrate is quickly released by the enzyme, which directs a water molecule into position to complete the hydrolytic process. Chymotrypsin enhances the rate of hydrolysis of peptide bonds by about 10^8–10^{10} compared to the rate in the absence of the enzyme.

This activity exhibits several of the important characteristics of enzyme catalysis:

1. Substrate and reaction specificity
2. Marked rate of enhancement of chemical reaction
3. A bound transition state that lowers the energy of activation of a reaction intermediate
4. Regeneration of the enzyme at the end of the reaction
5. A requirement for mild reaction conditions.

These characteristics depend upon the conformation of the protein, which in turn is dictated by its primary amino acid sequence. A protein, in solution, will fold to minimize its free energy. Hence, to understand enzyme catalysis and to be able to redesign enzymes for specific purposes, we must consider the forces that stabilize an enzyme in its particular conformation and yet permit some degree of molecular motion resulting in catalysis.

Quantitatively one of the most important of these stabilizing forces is the hydrophobic interaction. The removal of hydrophobic residues from the aqueous medium by folding into the interior of the protein leads to a decrease in the standard free energy. Hydrophobic interactions are entropically driven in that their removal into the interior of the protein increases the entropy of the surrounding water. Thus, these interactions are destabilized at lower temperatures but are stabilized by increasing the temperature over a modest range. Therefore, one possible method for minimizing the denaturation of an enzyme at an elevated temperature might be to increase the number of its hydrophobic interactions. Conversely, in a nonpolar solvent an enzyme would tend to unfold to expose its hydrophobic interior. Therefore, a decrease in the number of hydrophobic interactions might help in stabilizing an enzyme in organic solvents.

The relative importance of ionic interactions between ionized amino acid side chains of opposite polarity to the overall stabilization of protein structure is difficult to assess. Ionic interactions, like the hydrophobic interactions, are also entropically driven. This is perhaps somewhat surprising since it might be expected that the interactions are entropically unfavorable due to the decreased translational freedom of ionically bonded groups. However, in the unbonded state, the ions are heavily solvated by ion–dipole interactions with water molecules. These water molecules are liberated and become more disordered when ionic bonds are formed. In nonpolar solvents, ionic interactions should be stabilized because competing water molecules have been removed. However, changing the dielectric constant of the solvent or changing the pH of an aqueous solvent will have an effect on the ionization of the polar groups in the active site of the enzyme, which may have a profound influence on catalytic activity. Increasing the number of ionic interactions in an enzyme should help stabilize it against thermal denaturation.

A third force that contributes to the determination and maintenance of protein structure is hydrogen bonding. Hydrogen bonds can be formed by

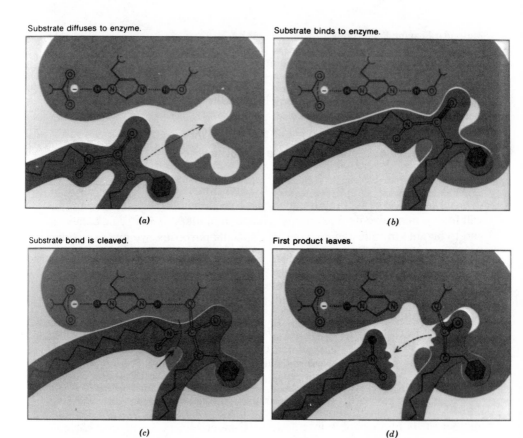

Substrate diffuses to enzyme.

Substrate binds to enzyme.

(a)

(b)

Substrate bond is cleaved.

First product leaves.

(c)

(d)

Figure 6.2. Schematic representation of the catalytic activity of the proteolytic enzyme chymotrypsin. The polypeptide chain to be cleaved binds first to the active site of the enzyme and inserts the proper side chain into the specificity pocket, which contains the functionally important residues Asp-102, His-57, and Ser-195 linked in a charge relay system (*a*). The histidine acts first as a general base pulling the serine proton toward itself and then as a general acid donating the proton toward the lone electron pair on the nitrogen atoms of the polypeptide bond to be cleaved (*b*). The aspartic acid helps the histidine attract the proton by making an electrostatic linkage with the other proton on the imidazole ring (*c*). As the serine H—O bond is broken, a bond is formed between the serine oxygen and the carbonyl carbon on the polypeptide chain. This carbon becomes tetrahedrally bonded, and the effect of the negative charge on Asp-102 has, in a sense, been relayed to the carbonyl oxygen atom of the substrate. This negative oxygen is stabilized by hydrogen bonds to the N—H groups on the enzyme backbone. However, as the polypeptide N accepts the proton from the histidine, the N—C bond is weakened and finally broken (*d*).

Water molecule approaches.   Water reacts with acyl enzyme.

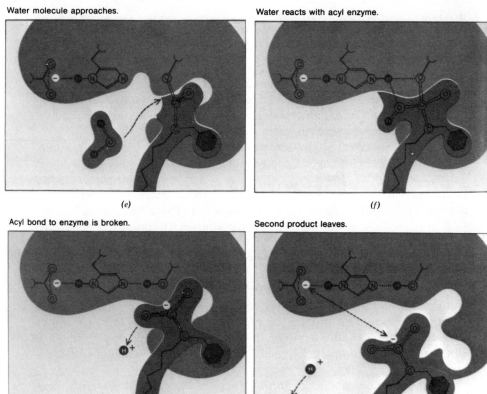

(e)   (f)

Acyl bond to enzyme is broken.   Second product leaves.

(g)   (h)

Figure 6.2 (*continued*). One half of the polypeptide chain falls away as a free amine, R—NH$_2$. The other half remains covalently bonded to the enzyme as an acyl intermediate. The deacylation of the enzyme is similar to the first few steps in reverse, with H$_2$O playing the role of the missing half-chain (*e,f*). A water molecule attacks the carbonyl carbon of the acyl group and donates one proton to His-57 to form another tetrahedral intermediate. This breaks down when the proton is passed from the histidine to serine (*g*). Then the other half of the polypeptide chain falls away, aided by charge repulsion between the carboyl group and the negative charge on Asp-102, and the enzyme is restored to its original state (*h*). (G. Zubay, Biochemistry, © 1983, Addison-Wesley Publishing Company, Inc., Reading, MA. Pgs 150–151, Fig 4-10. Reprinted with permission.)

the C=O and N—H groups of each peptide bond and the electronegative atoms of polar side groups either between themselves or with water. In the interior of the protein, hydrogen bonds between the amide nitrogen and carbonyl oxygen of the backbone predominate. These hydrogen bonds have a major role in stabilizing the ordered secondary structures, the α-helix and the β-sheet, and in stabilizing the interactions between these ordered structures.

Forces that are especially important in holding protein subunits together are van der Waals attractions. Unlike the other forces discussed above, which contribute 3-7 kcal/mol, van der Waals are weak, each accounting for only a few hundred calories. However, over an extensive surface van der

Waals interactions can contribute 10–50 kcal toward the energy of subunit association.

Finally proteins are stabilized by covalent disulfide bonds between juxtaposed cysteine residues. These bonds contribute about 60 kcal/bond and thus are very important in maintaining protein structure. Addition of a cysteine pair to an enzyme should lead to an improvement in overall stability, but the pair would have to be located carefully in order to stabilize a critical component without interfering with catalytic activity.

Taking into account all the energies that contribute to the stabilization of protein structure, it might be supposed that an enzyme is quite stable under normal conditions. However, it sustains a severe entropic penalty when folding into a constrained structure, and at room temperature parts of an enzyme are usually quite close to denaturation. These "hot spots" of unfolding might be a good place to modify a protein in order to improve its stability against thermal or chemical denaturation.

Enzymes are frequently considered to be unnecessarily complicated. Parts of the many enzymes are thought to be involved in functions that are only important for activity inside a cell, for example, regulation, transport, and processing. Such functions might not be required when the enzyme in question is being used in vitro for organic synthesis. Can these unwanted parts of the molecule be removed or replaced via genetic engineering and will these modifications be beneficial to the activity or stability of the enzyme?

A consideration of the chemical principles that govern protein folding helps to focus attention on the critical part of a protein that requires modification for some specific application. It also provides the rational framework for choosing the most likely amino acid changes to accomplish the desired modification. Unfortunately, the principles are not sufficiently understood to permit the changes to be made with a high degree of certainty as to the outcome. Indeed, one of the key benefits of recent experiments on enzyme modification by genetic engineering is the additional insight it has given into the relationship between amino acid sequence, protein structure, and enzyme mechanisms. This will be discussed in the next section.

3. MODIFICATION OF PROTEINS BY SITE-DIRECTED MUTAGENESIS

3.1. Introduction

Mutagenesis is the process of altering the genetic material of an organism, usually by the application of certain chemicals or by irradiation with UV light or X-rays. Mutagenesis has been used for several decades as a tool for dissecting gene function and for altering enzyme activity. But there are several drawbacks to classical mutagenesis. First, only random mutations can be introduced into the genetic material. Therefore, only very rarely, and by chance, does a mutation occur in the gene of interest. Even when such a mutation occurs, it is frequently detrimental to the organism. Generally, only a mutation that results in a novel activity that provides an advantage for the organism can be selected for. In the absence of a selection technique it is extremely laborious, and frequently impossible, to isolate a mutant strain.

The possible occurrence of mutations other than the one being selected is a second disadvantage of classical mutagenesis. Such mutations might escape initial detection but could subsequently be found to be undesirable. A third disadvantage is that it is almost impossible to select simultaneously for two or more desirable mutations, for example, two amino acid changes in an enzyme. Therefore, major restructuring of enzymes is not possible using classical mutagenesis.

In contrast, site-directed mutagenesis permits one or more selected amino acids in a specific enzyme to be precisely altered or replaced, eliminating the problems inherent in classical mutagenesis. Site-directed mutagenesis has only recently become possible due to advances in genetic engineering. It requires the ability to:

1. Isolate the appropriate gene encoding the desired enzyme using recombinant DNA techniques (as discussed in the previous chapter).
2. Alter the gene at precisely the correct nucleotide(s).
3. Reintroduce the mutated gene back into an organism in a construction that expresses the mutated gene as an altered protein. This is accomplished by standard recombinant DNA technique as discussed in the previous chapter.

This section provides three examples that illustrate different capabilities of the technique of site-directed mutagenesis:

1. The ability to alter a single DNA base resulting in a specified amino acid change (tyrosyl tRNA synthetase).
2. The ability to replace a segment of DNA in a gene in order to facilitate subsequent mutation, the so-called cassette approach (subtilisin BPN').
3. The ability to synthesize a complete gene, introducing localized mutations and facilitating subsequent mutations (ribonuclease S).

3.2. Tyrosyl tRNA Synthetase

The enzyme tRNA synthetase catalyzes the addition of an amino acid to its cognate tRNA. Catalysis is carried out in two steps. In the first, or activation, step an aminoacyl AMP is synthesized in a reaction between an amino acid and ATP:

$$^+H_3N-\underset{\underset{R}{\mid}}{\overset{\overset{H}{\mid}}{C}}-\overset{\overset{O}{\parallel}}{C}-O^- + ATP \rightleftarrows H_3N-\underset{\underset{R}{\mid}}{\overset{\overset{H}{\mid}}{C}}-\overset{\overset{O}{\parallel}}{C}-\underset{\underset{O}{\mid}}{O}-\overset{\overset{O}{\parallel}}{P}-Ribose-AMP + PP_1$$

Pyrophosphate

In the second, or transfer, step the aminoacyl-AMP reacts with tRNA to form acylated tRNA and AMP. The acylated tRNA is said to be charged and can now participate in placement of the loaded amino acid into a growing peptide at the site of cellular protein synthesis (the ribosome).

Site-directed mutagenesis has been used to improve the affinity of tyrosyl tRNA synthetase for its ATP substrate by about 100-fold [Wilkinson et al., 1984]. It had previously been speculated by inspection of the x-ray crystallographic model of the enzyme that the side-chain hydroxyl of the threonine-51 participated in a weak hydrogen bond with the AMP moeity of the substrate intermediate. However, in the absence of substrate it was predicted that the hydroxyl group would make a strong hydrogen bond with water, which would be expected to favor dissociation of the enzyme-substrate complex. Two independent point mutations were constructed with a view to eliminating the hydroxyl side chain, $Thr_{51} \rightarrow Ala_{51}$ and $Thr_{51} \rightarrow Pro_{51}$. The Pro_{51} mutation also resulted in a distortion of the polypeptide backbone. Both mutants have increased activity for ATP. The Pro_{51} mutation resulted in a 25-fold increase mainly due to a lowering of the K_m for ATP. K_m is a measure of the binding of the substrate to the enzyme. A low K_m value means that the enzyme reaches its maximum catalytic rate at a lower concentration of substrate and, generally, indicates that the enzyme binds its substrates more tightly.

How was the Thr_{51} precisely altered to Ala_{51}? First, the entire gene for tyrosyl-tRNA synthetase was cloned into the single-stranded DNA vector, M13, by techniques described in Chapter 5. Since the gene had previously been sequenced, it was possible to identify the region corresponding to the thr_{51} position. Next, a short oligonucleotide was synthesized chemically that corresponded exactly to the complimentary DNA region around thr_{51} but containing the codon GCG (Ala) in place of the naturally occurring AGC (Thr) codon. The steps that result in the incorporation of the synthetic oligonucleotide in the synthetase gene are outlined schematically in figure 6.3. The end result is a population of bacteria containing either the wild-type or mutant gene. These can be distinguished by hybridization of the synthetic oligonucleotide to the DNA of the bacteria under conditions that strongly favor the formation of perfectly matched double strands of DNA. These conditions result in the visible hybridization of the synthetic oligonucleotide

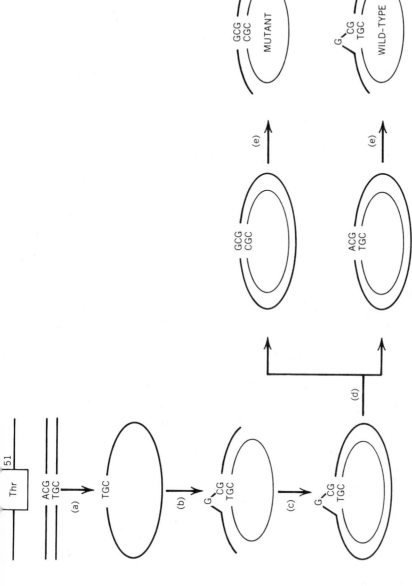

Figure 6.3. General strategy for site-directed mutagenesis. (*a*) A gene (in this case the gene for tyrosyl tRNA) is cloned into the single-stranded DNA vector M13. (*b*) An oligonucleotide of about 18 residues is chemically synthesized. The oligonucleotide is complementary to a region of the gene that contains the coding sequence to be mutated. The oligonucleotide contains one mismatched nucleotide that will result in the desired mutation. (*c*) A complete double-stranded vector is synthesized in vitro using enzymes to extend the synthetic oligonucleotide and to close the circle. (*d*) The double-stranded vector is introduced into a bacterial cell that will produce replicas of the vector including the mutation. (*e*) The bacterial cells harboring the mutant gene can be detected by hybridization with the synthetic oligonucleotide.

to the mutated gene and, thus, the bacteria harboring the mutant form of the gene can be detected and isolated.

It is clear that this strategy could only have been accomplished with the advent of genetic engineering techniques that permit the identification and isolation of individual genes. It is also obvious that the ability to sequence DNA and to synthesize chemically short regions of genes plays a central role in the technique of in vitro mutagenesis. One other key requirement is a knowledge of the crystal structure of the enzyme. Without this knowledge it would be hard, or impossible, to select the appropriate mutations that could potentially alter the properties of an enzyme in a desirable way.

3.3. Subtilisin BPN'

In the example discussed above, two independent point mutations were made at a site that was correctly predicted to be important for substrate binding. These two point mutations were introduced by a basic strategy that must be repeated for each new amino acid change that is desired. Since the procedure entails tedious and, in some cases, inefficient steps, an alternate procedure has been developed that facilitates multiple amino acid changes to be made in an enzyme. This alternate procedure is termed the "cassette" approach since it readily permits a segment of the gene to be removed and replaced with any other desired segment. Initially, the basic site-directed mutagenesis procedure described above is used to introduce restriction-enzyme sites on either side of a region to be mutated. The introduction of these sites then permits that region to be excised and to be replaced with chemically synthesized DNA containing the desired mutation. The advantages of this technique are that it permits the incorporation of more than one mutation simultaneously and it facilitates the incorporation of single mutations at different positions but in successive experiments. The "cassette" can be readily pulled out, modified, and replaced. An example of this approach is the modification of subtilisin BPN' (Estell et al., 1984).

Subtilisin BPN' is a serine protease excreted by *Bacillus amyloliquefaciens*. Its amino acid sequence, three-dimensional structure, and reaction mechanisms are known. The enzyme has a large crevice, the S1 site, which shows a distinct preference for binding hydrophobic or aromatic amino acids. This crevice is formed from three segments of the molecule. One of these segments, Val_{165}-Gly_{166}-Tyr_{167}-Pro_{168}, forms the top of the crevice. It was expected that the side chain of an amino acid at position 166 would protrude into the crevice. Therefore replacement of Gly_{166} might alter substrate specificity.

In order to investigate the effects on substrate specificity Estell and co-workers (1984) have replaced Gly_{166} with all possible amino acids. The strategy to achieve this is illustrated in Figure 6.4. First, suitable restriction sites were created by site-directed mutagenesis at two sites closely flanking the target codon on the gene for subtilisin-BPN' (Fig. 6.4a). These restriction

<pre>
                                    160   166
Wild-type amino acid sequence: thr ser gly ser ser ser thr val gly tyr pro gly
     Wild-type DNA sequence: 5' ACT TCC GGC AGC TCA AGC ACA GTC GGC TAC CCT GGT 3'

                    a. 5' ACT TCC GGG*AGC TCA AGC ACA GTC GGC TAC CCG*GGT 3'
                                  SacI   XmaI

                    b. 5' ACT TCC GGG AGC T   pCCG GGT 3'
                           TGA AGG CCCp   CA

                    c.   CA AGC ACA GTC NNN TAC
                                            TCG AGT TCG TGT CAG NNN ATG GGC C
</pre>

Figure 6.4. Strategy for the mutagenesis of Gly$_{166}$ to any amino acid in the protein subtilisn-BPN'. (*a*) The first step requires the creation of two unique restriction enzyme sites, SacI and XmaI, that span the Gly$_{166}$ position. (*b*) These two restriction enzymes are used sequentially to remove the intervening segment, (*c*) which can then be replaced with a synthetic DNA fragment (the so-called cassette) containing any codon at the Gly$_{166}$ position. N = any nucleotide. * = mutagenized nucleotide. (Estell et al., 1984. Redrawn with permission.)

sites were designed to conserve the amino acid sequence at these positions. The segment between the two new restriction sites was excised (Fig. 6.4*b*) and replaced, in turn, with short chemically synthesized segments of DNA that contained the code of all 19 possible amino acids to replace Gly$_{166}$ (Fig. 6.4*c*).

Two substitutions, involving negatively charged Asp$_{166}$ and Glu$_{166}$, might be expected to favor substrates with positively charged amino acids. This was found to be the case. Using a substrate that could correctly load a phenylalanine in the S1 site, it was found that substitution of Asn or Gln at position 166 had little effect compared to the wild-type. However with substitution of Asp or Glu at position 166, the K_{cat}/K_m ratio dropped by a factor of 100 and 10, respectively. The situation is quite different when the substrate contains arginine instead of phenylalanine; then the Asp and Glu substitutions have K_{cat}/K_m ratios 25 and 35 times greater than wild-type.

3.4. Ribonuclease S

In the example described above a segment of the subtilisin-BPN' gene was modified by incorporation of specified restriction enzyme sites so that short in vitro synthesized oligonucleotides could be readily inserted between the restriction sites. An extension of this strategy that we will now discuss permits even greater flexibility of design. The idea is to chemically synthesize an entire gene. Codons can be chosen which are consistent with the desired amino acid sequence, but which simultaneously incorporate strategically located restriction enzyme sites. The sites can be used to delete,

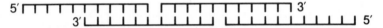

Figure 6.5. Strategy for the construction of a synthetic gene. Short complementary and over-lapping oligonucleotides of about 20 residues are synthesized. They are permitted to reas-sociate, and the gaps between the fragments are covalently closed with ligase. The intact segment is then cloned into a convenient expression vector as described in Chapter 5.

insert, or alter the amino acid sequence anywhere in the gene. In principle such sites could be located throughout the gene provided that they do not interfere seriously with the efficiency of transcription or translation.

The problem with this approach is that it is currently not practical to synthesize oligonucleotides of lengths sufficient to encode enzymes. This would require, typically, the synthesis of a DNA fragment greater than 300 base pairs. Currently it is feasible to construct oligonucleotides up to 40 residues long with reasonable yields. However, the problem can be circum-vented by synthesizing short (about 20 nucleotides) complementary and overlapping DNA segments. These are then permitted to reassociate to form an intact gene as illustrated in Figure 6.5. Using this technique a full length gene for ribonuclease S has been synthesized and expressed in *E. coli* (Nabir et al., 1984). The gene, consisting of 330 base pairs, was constructed from 66 different oligonucleotides of 10–22 residues in length. The advantage of this strategy is that it permits the subsequent rapid modification by replacement with synthetic duplex DNA of any section of a gene that is located between the various restriction enzyme sites previously built into the gene.

At the present time, the synthetic gene for ribonuclease S has been cloned and used to synthesize an authentic protein. It has not yet been subjected to further modification. However, one can consider the sorts of modification that might be constructive and illuminating given the ability to modify any part of the protein. For these considerations the ribonuclease S protein seems to be a particularly good candidate. Six crystal structures of ribonu-clease and its variants are available, including three-dimensional structures of ribonuclease bound to reactants, products, and transition state analogues. The primary amino acid sequences of over 40 mammalian ribonucleases are known. These sequences can be compared and used to indicate positions in which alterations are interesting. There is also a considerable amount of biochemical and chemical information available and methods of kinetic and thermodynamic analysis make investigation of catalysis by ribonuclease subject to rigorous physical and organic chemistry.

In addition to being well suited to the study of catalysis, ribonuclease is appropriate for the study of many other important problems in biological chemistry. Ribonuclease is a model enzyme illustrating virtually every step in protein folding. The aggregation between the S peptide and the S protein is an excellent model for chain association and has been studied by system-atic alteration of S peptides. Ribonuclease is believed to have a hydrophobic nucleation site for folding. Cis-trans isomerization of proline is believed to determine the rate of folding. Ribonuclease is a convenient system for studying the formation of disulfide bonds. Furthermore, the protein is suited for detailed studies on the chemical nature of thermal stability of proteins. Each one of these important questions, involving structure-function relationships, can be addressed by systematically altering the protein by site-directed mutagenesis.

3.5. Enzymatic Properties Amenable to Modification

An important aspect of site-specific mutagenesis is its use as a tool for elucidating the relationship between the structure of an enzyme and its function. Increasing knowledge in this area will facilitate future attempts to introduce desired alterations into enzymes. Currently, modifications such as the enhancement of activity (tyrosyl-tRNA synthetase) or altered substrate specificity (subtilisin BPN') are accomplished by educated guesswork based on a knowledge of the crystal structure, reaction mechanism, and amino acid sequence of the enzyme in question. In the future the additional knowledge of structure and function relationships provided largely by advances in site-specific mutagenesis will permit more effective modifications to be made. Some day it may permit the design of totally novel enzymes not found in nature.

Given these prospects one can consider the sorts of modifications that it would be desirable to control in a predictable fashion (Ulmer, 1983). These might include;

1. Kinetic properties, including the turnover number of the enzyme and the Michaelis constant for a particular substrate.
2. Thermostability and temperature optimum.
3. Stability and activity in nonaqueous solvents.
4. Substrate and reaction specificity.
5. Cofactor requirements.
6. pH optimum.
7. Protease resistance.
8. Allosteric regulation.
9. Molecular weight and subunit structure.
10. Molecular handles, such as the incorporation of reactive amino acids that would permit specific attachment of the enzyme to a solid support or for the attachment of additional reactive chemical groups.
11. Incorporation of unnatural amino acids.

The rapid progress in this area is largely due to advances in genetic engineering that permit the isolation and manipulation of individual genes. Cloned genes are also a source for the rapid determination of amino acid sequences, since it is easier to sequence the DNA of the gene than the protein itself. To date, over six million bases of DNA sequence have been compiled and are now available to the public in the various Nucleic Acid Data Banks (W. Rinbone, personal communication, 1986). But, in addition to information on the amino acid sequence, the three-dimensional crystal structure of an enzyme is highly desirable. The determination of the structure of the enzyme by x-ray crystallography is currently a limiting step in the advancement of enzyme engineering. However, major advances have been made in the collection and analysis of diffraction data that permit the more rapid analysis of protein structure. These include the use of:

1. Synchrotron X-ray sources that reduce data collection time.
2. Position-sensitive X-ray detection for recording data that also reduces data collection time and simplifies subsequent processing steps.
3. Better algorithms that facilitate structure determinations.
4. Improved computer graphics that permit better visualization and manipulation of the three-dimensional structure.

It has also been suggested that, in some cases, the inspection of three-dimensional structures might not be necessary for redesigning enzymes (Orr, 1984). Nature has been manipulating enzyme structures for millions of years. Some information on the relationship between enzyme structure and function could be obtained by comparing the same enzyme from many different organisms and correlating the various sequences with the observed changes in the properties of the enzyme. For example, as we described above, the amino acid sequences of 40 different ribonucleases are now known. Nearly 70% of the amino acids in the enzyme have been altered as the organisms evolved. These alterations might, to some extent, be correlated with variations in substrate specificity, catalytic rate, and enzyme stability. If such a correlation is possible, it might prove useful in predicting the outcome of further modifications.

4. CHEMICAL MODIFICATION OF PROTEINS

4.1. Introduction

In the previous section we explored the ways genetic engineering can be used to assist in understanding how enzymes work and, in some cases, even improving their properties for specific applications. A very different approach, and a much older one, is to alter enzymes by direct chemical modification.

Proteins are susceptible, of course, to many chemical reactions. Most of these reactions are too difficult to control to result in specific modifications of interest without permitting a number of other changes in the enzyme to occur. Furthermore, certain interesting modifications cannot be carried out, such as those on the interior of the protein, which is often inaccessible to chemical reagents. Despite these limitations, several important specific modifications have recently been carried out. We will discuss some of these in order to demonstrate the power and versatility of chemical modification.

4.2. Myoglobin

Harry Gray and colleagues are interested in determining the extent to which the oxidation–reduction properties of a metalloprotein can be manipulated by the attachment of a redox-active inorganic group to a polypeptide chain ligand (Margalit et al., 1983). In one experiment, Gray and his colleagues attached a ruthenium electron transfer catalyst $[(RuNH_3)_5]^{3+}$ to three surface histidines of sperm whale myoglobin to produce a "semisynthetic biorganic enzyme." The function of myoglobin is to bind and transport oxygen especially in muscle tissue. However, the myoglobin modified with the ruthenium catalyst is able to reduce oxygen while oxidizing various organic substrates such as ascorbate and durohydroquinone. Comparison of the catalytic parameters (K_m, K_{cat}) indicates that the presence of the dioxygen binding site of the coordinated heme in the holomyoglobin is required for this activity (see Table 6.1). In the case of ascorbate, the rate of oxidation is enhanced approximately 200-fold and approaches that achieved by ascorbate oxidase. Gray speculates that in view of the proximity of His-113 to the heme, it is likely that a Ru(His-113) transfers an electron rapidly to the heme–dioxygen complex, thereby producing some form of heme-bound peroxide intermediate (whose dissociation may prove to be the rate-limiting step).

4.3. Papain

Another dramatic way in which the function of a protein can be changed by chemical modification is exemplified by the work of Emil Kaiser's laboratory at Chicago University and the Rockefeller University. They modified

TABLE 6.1 Reactivity Parameters for Oxidations Catalyzed by Ruthenium, Myoglobin, and Related Species at 25°C

Substrate	Catalyst[a]	K_m, M^{-1}	K_{cat}/S
Ascorbate	RuMb	1.5×10	0.6
	RuapoMb	1.9×10	0.0063
	aRuIm	1.3×10	0.0035
Durohydroquinone	RuMb	2.8×10	0.3
	RuapoMb	1.1×10	0.058
Hydroquinone	RuMb	1.1×10	0.0042

Source: Margalit et al., 1983.
[a] aRuIm, pentaamineruthenium (III); RuMb, holomyoglobin modified with ruthenium catalyst; RuapoMb, apomyoglobin (lacking coordinated heme) modified with ruthenium catalyst.

the proteolytic enzyme, papain, by attachment of a synthetic flavin, an oxidation–reduction catalyst (Slama et al., 1981; Kaiser et al., 1980). This converted a hydrolytic enzyme to one capable of carrying out oxidations. X-ray structure information and solution data indicated that it might be possible to introduce a functional group into the active site of papain and still leave room for substrate binding. In addition, an easily reacted cysteine residue within the active site of papain could be activated specifically and used to bond the synthetic functional group to the enzyme covalently.

A flavin reagent seemed an optimum choice for the synthetic part of the new molecule because this reagent seems to make relatively few demands on the enzymes that it is normally associated with. Thus, even a poor match between flavin and papain would be expected to show some activity.

Several synthetic flavins were constructed and attached to the papain, and their abilities to oxidize N'-alkyl-1,4-dihydronicotinamides were compared to those of the flavins alone or the naturally occurring nicotinamide adenine dinucleotide (NADH) dehydrogenase. The flavopapain prepared from a 7-bromo-acetyl substituted flavin (compound 1, Fig. 6.6) gave a rate increase, compared to flavin alone, of one to two orders of magnitude depending on the substrate. The material also showed saturation kinetics characteristic of an enzyme reaction. The flavopapain produced from an 8-bromoacetyl-substituted flavin (compound 2, Fig. 6.6) gave the best results. This material not only displays saturation kinetics, but it gives a thousandfold increase in rate compared to model reactions. Compound 2 is the most efficient semisynthetic enzyme constructed to date and approaches the activity displayed by all but the most efficient flavin-containing oxidoreductases known.

4.4. Polyethylene Glycol Adducts

An important class of compounds has been produced by addition of polyethylene glycol (PEG) or polyethylene oxide (PEO) to enzymes. This addition can result in improved solubility, especially in organic solvents, better stability in aqueous/organic solvent mixtures, and better resistance to proteolytic degradation. For example, three or four PEG molecules have been added to the enzyme lipase (Inada et al., 1984). Lipase normally catalyzes

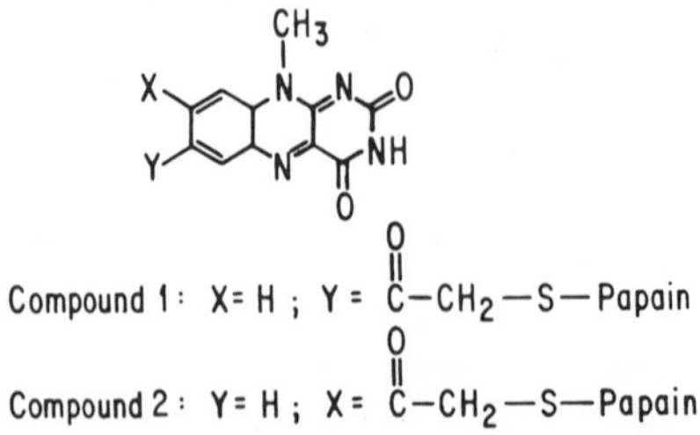

Compound 1 : X= H ; Y = C−CH$_2$ —S—Papain

Compound 2 : Y= H ; X = C−CH$_2$ —S—Papain

Figure 6.6. Structure of two flavopapains. (Slama et al., 1981. Reprinted with permission.)

TABLE 6.2. Activity of Ester Synthesis from Pentyl Alcohol (0.75 M) and Pentanoic Acid (0.50 M) Catalyzed by Modified Lipase in Various Organic Solvents at 20°C

Organic Solvent[a]	Activity (μmol/min/mg of protein)
Benzene (water saturated)	1.82
Benzene	0.84
Toluene	0.73
Chloroform	0.56
Dioxane	0.19
Dichloromethane	0.07
Dimethylformamide	0.05
Acetone	0.00
Dimethoxyyethane	0.00

Source: Inada et al., 1984.
[a] Each solvent was dehydrated with molecular sieves 3A, except the water-saturated benzene.

the hydrolysis of triglycerides to glycerol and fatty acids. However, it was found that the PEG-bound lipase was soluble in various organic solvents and in these solvents was able to catalyze the esterification of a number of substrates such as pentyl alcohol to pentyl pentanate, lauryl alcohol + stearic acid to lauryl stearate, and cholesterol + stearic acid to cholesterol stearate + H_2O + sitosterol. These reactions are the reverse of the normal activity of lipase.

The activity of modified lipase in the synthesis of pentyl pentanate in various organic solvents was explored. As can be seen from Table 6.2 its activity is most pronounced in water-saturated benzene (30 mM H_2O), but it is also active in benzene, toluene, and chloroform and to a lesser degree in several other solvents. In addition to ester synthesis, the lipase was demonstrated to be able to carry out ester exchange and aminolysis in benzene.

Thermolysin is another example that dramatically illustrates, the alteration of enzymatic activity after modification with PEG. Thermolysin is normally a protease catalyzing the cleavage of peptide bonds in aqueous media. However, after modification with PEG, it is capable of synthesizing peptide bonds in organic solvent and has been used to make a precursor of the artificial sweetener, aspartame (the methyl ester of asparatic acid linked to phenylalanine) (Nakanishi et al., 1985). Thermolysin cleaves peptide bonds in living cells by addition of water across the peptide bonds. But, like all enzymes, the forward reaction (hydrolysis) is in equilibrium with the reverse reaction (synthesis). In water, the reaction rate of the synthesis is very low compared with that of the hydrolysis, but this equilibrium can be shifted by replacement of the bulk water phase with an organic solvent. The fact that the modified thermolysin has catalytic activity in an organic solvent may be startling. There are 15 known proteases with different specificities for peptide cleavage; therefore, it might be possible to synthesize a peptide in vitro under conditions that reverse the cleavage but maintain the specificity of the proteases.

4.5. Lactate Dehydrogenase

Finally, in this section we will describe an example of a more general chemical modification of a protein, namely the acetamidination of lactate dehydrogenase (Tuengler and Pfeiderer, 1977). Lactate dehydrogenase is a tetrameric protein with a molecular weight of 144,000. It catalyzes the reduction of pyruvate to lactate. The reaction requires NAD^+. Each subunit has 24 lysines, 17 of which appear to be on the surface. No lysine is involved in catalysis.

Acetamidination is a mild procedure that is specific for the ε-amino groups of lysines without altering the charge:

$$H_3C-C\overset{NH_2^+}{\underset{OCH_3}{\diagdown}} \quad + \quad H_2N-(CH_2)_4-\overset{\overset{|}{NH}}{\underset{\underset{|}{C=O}}{CH}} \rightarrow$$

$$\overset{H_2N}{\underset{H_3C}{\diagup}}C-NH-(CH_2)_4-\overset{\overset{|}{NH}}{\underset{\underset{|}{C=O}}{CH}} + CH_3OH$$

The acetamidination is restricted to the 17 surface lysines and appears to have little effect on the catalytic activity. However, it does enhance the stability of the enzyme toward heat and alkali denaturation as well as tryptic digestion. It has been proposed that the increase in thermal stability may result from the arginine-like character of the derivitized lysines. The side chain of arginine is by far the most hydrophilic of the residues. When it is placed on the surface of the protein, it is highly solvated and resists withdrawal from the aqueous surroundings during denaturation. Interestingly, thermostable lactate dehydrogenase from thermophilic bacteria differs from the mesophilic lactate dehydrogenase by an increase in the arginine to lysine ratio on the surface of the protein.

5. NOVEL CATALYTIC ACTIVITIES OF ENZYMES

5.1. Introduction

In living organisms, enzymes carry out thousands of biochemical reactions. Although no system of classification of enzymes or their reactions is perfect, one of the favored schemes adopted by The International Union of Biochemistry is illustrated in Table 6.3. Many of these enzyme-catalyzed reactions are of potential importance to the synthetic chemist. For example, the reactions carried out by the oxidoreductases and the lyases are among the keystones of synthetic organic chemistry. The hydrolases catalyze reactions that are of considerable general synthetic value for selective hydrolysis and in resolution, particularly for esters of racemic acids. However, these and other enzymes are generally most soluble and stable in water or in water containing relatively small quantities of polar co-solvent such as polyhedric alcohols or dimethyl sulfoxide. Also, enzymes normally function best with substrates that are soluble in these media. On the other hand, organic synthesis, in recent years, has focused on terpenes, steroids, alkaloids, prostanoids, and other classes of water-insoluble substances.

The potential value of using enzyme catalysis for different or possibly novel organic synthesis problems has long been recognized. The generally high catalytic efficiencies of enzymes, some of which enhance reaction rates 10 billionfold is an attraction. Furthermore, the reactions are catalyzed under very mild conditions, often at 20°C and pH 7.0. This enables problems, such as isomerization, epimerization, racemization, and rearrangement, encountered with many organic reactions on sensitive molecules to be avoided. However, the selectivities of reactions achievable and above all the stereospecificity with which they can be effected are the most important features of enzymatic catalysts from the organic chemist's point of view.

For these reasons, a considerable effort is underway to discover novel enzymes with unusual catalytic properties and, especially, to explore the catalytic activity of known enzymes with novel substrates or under unusual reaction conditions.

TABLE 6.3. Enzyme Reaction Types

1. Oxidoreductases	$C-H \rightarrow C-OH$
	$CH(OH) \rightleftharpoons C=O$
	$CH-CH \rightleftharpoons C=C$
2. Transferases	Transfer aldehyde, ketone, acyl, sugar, and phosphoryl groups
3. Hydrolases	Hydrolysis of broad range of substrates
4. Lyases	Addition or formation of double bonds, e.g., $C=C$, $C=O$, $C=N$
5. Isomerases	Various isomerizations including racemization
6. Ligases	Formation of $C-O$, $C-S$, $C-N$ bonds

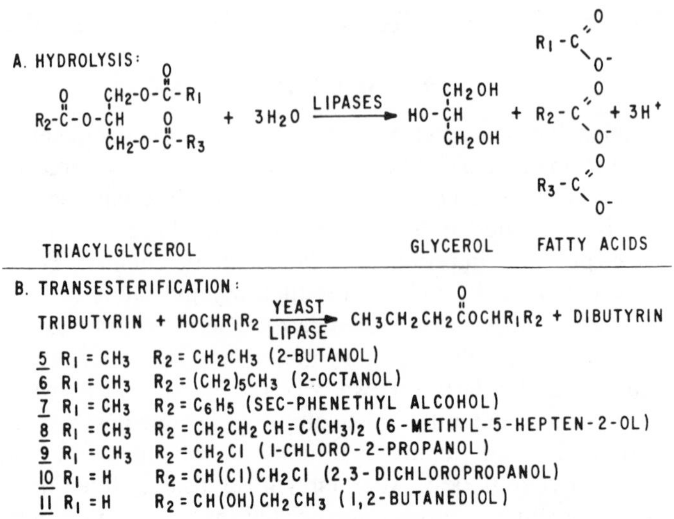

A. HYDROLYSIS:

$$
\begin{array}{c}
\text{O} \quad \text{CH}_2\text{-O-C-R}_1 \\
\text{R}_2\text{-C-O-CH} \quad \text{O} \\
\text{CH}_2\text{-O-C-R}_3
\end{array}
+ 3H_2O \xrightarrow{\text{LIPASES}}
\begin{array}{c}
\text{CH}_2\text{OH} \\
\text{HO-CH} \\
\text{CH}_2\text{OH}
\end{array}
+ R_2\text{-C}\overset{O}{\underset{O^-}{\diagdown}} + 3H^+
$$

$$R_1\text{-C}\overset{O}{\underset{O^-}{\diagdown}}$$

$$R_3\text{-C}\overset{O}{\underset{O^-}{\diagdown}}$$

TRIACYLGLYCEROL   GLYCEROL   FATTY ACIDS

B. TRANSESTERIFICATION:

$$\text{TRIBUTYRIN} + \text{HOCHR}_1\text{R}_2 \xrightarrow[\text{LIPASE}]{\text{YEAST}} \text{CH}_3\text{CH}_2\text{CH}_2\overset{O}{\text{C}}\text{OCHR}_1\text{R}_2 + \text{DIBUTYRIN}$$

5 $R_1 = CH_3$ $R_2 = CH_2CH_3$ (2-BUTANOL)
6 $R_1 = CH_3$ $R_2 = (CH_2)_5CH_3$ (2-OCTANOL)
7 $R_1 = CH_3$ $R_2 = C_6H_5$ (SEC-PHENETHYL ALCOHOL)
8 $R_1 = CH_3$ $R_2 = CH_2CH_2CH=C(CH_3)_2$ (6-METHYL-5-HEPTEN-2-OL)
9 $R_1 = CH_3$ $R_2 = CH_2Cl$ (1-CHLORO-2-PROPANOL)
10 $R_1 = H$ $R_2 = CH(Cl)CH_2Cl$ (2,3-DICHLOROPROPANOL)
11 $R_1 = H$ $R_2 = CH(OH)CH_2CH_3$ (1,2-BUTANEDIOL)

Figure 6.7. (*a*) The naturally occurring enzyme lipase cleaves triglycerol to glycerol and fatty acids. (*b*) Lipase can also catalyse the transesterification of tributyrin to various alcohols. (Cambou and Klibanov, 1984. Reprinted with permission.)

5.2. Lipase

The biological action of lipase is to hydrolytically cleave triglycerides to glycerol and fatty acids (Fig. 6.7*a*). However, it has been known for some time that lipase can carry out transesterification of various nucleophiles such as alcohols, but yields have been low due to competition with water. A certain amount of water is necessary to maintain the integrity of the enzyme, but in certain instances this can be made very low. Klibanov and his colleagues have investigated the conditions necessary to achieve transesterification of organic esters to various alcohols using lipases derived from hog liver or yeast (Cambou and Klibanov, 1984). In these experiments a porous support (sepharose or chromosorb) was filled with an aqueous solution of the enzyme and used to convert the organic substrate ester, tributyrin (tributyryl glycerol), to the various alcohols listed in Figure 6.7*b*. The advantages of using beads to retain the enzyme in the aqueous component are (1) the robustness of the bead, (2) the ability to separate them readily at the end of the reaction from reactants, and (3) the opportunity to reuse them repeatedly. A general scheme for the production of optically active alcohol by this technique is illustrated in Figure 6.8.

The advantage of this scheme is that water-insoluble substrates can constitute the organic phase, whereas the enzyme is located in the aqueous phase. Since the fraction of the latter phase can be made very low, such an arrangement solves the problem of both the competition of an alcohol (the nucleophile) with water in the enzymatic reaction and poor solubility of most organic esters (the substrates) and alcohols (the products) in water.

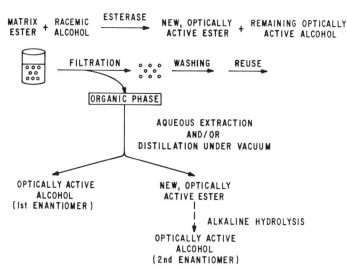

Figure 6.8. A general scheme for the production of optically active alcohols by the lipase-catalyzed transesterification of esters. (Cambou and Kilbanov, 1984. Reprinted with permission.)

Klibanov and his colleagues went one step further (Zaks and Klibanov, 1984). They were able to demonstrate that porcine lipase was active even when the concentration of water in a mixture of tributyrin and water was as low as 0.015%. In this medium the activity of the enzyme is five times greater at 100°C than at 20°C. They also find that dehydration changes the specificity of the enzyme. "This phenomenon, should it prove to be a general one," they note, "may form a basis for a new approach to the improvement of catalytic properties of enzymes."

5.3. Alcohol Dehydrogenase

Another example of an enzyme that catalyzes important reactions with a broad spectrum of "unnatural" substrates is alcohol dehydrogenase, which can oxidize a variety of acyclic, mono-, and bicyclic and tetracyclic (steroid) substrates. It operates on most of them with high enantiomeric specificity. It also exhibits prochiral stereospecificity. Prochiral stereospecificity of enzymes is extremely important since it permits asymmetric synthesis from symmetric starting materials. Furthermore, stereospecificity of alcohol dehydrogenase is predictable for both acyclic and cyclic substrates.

The work of Jones (1982) using horse liver alcohol dehydrogenase (HLADH) serves as a model study to illustrate several of these points. For example, HLADH can controllably synthesize R or S alcohols using bridged bicyclic racemic ketones as a starting material. The preparation of such stereoisomers is difficult chemically but can be acheived easily using HLADH.

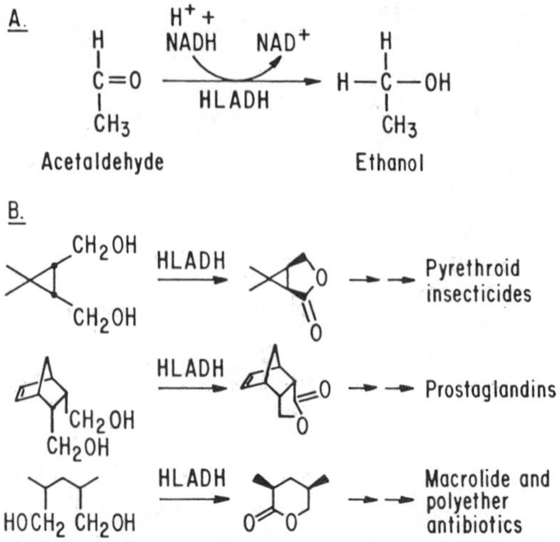

Figure 6.9. (*a*) The biological function of HLADH is the catalytic reduction of acetaldehyde to alcohol. (*b*) HLADH can oxidize a very broad range of mesodiols to give the corresponding lactone of 100% enantiomeric purity. Several of these chiral lactones are excellent starting materials for a number of useful products. (Jones, 1982. Redrawn with permission.)

A major advantage of enzyme catalysis is the ability to combine several different kinds of specificity and thereby achieve in one step an overall reaction requiring several separate chemical reactions. For example, HLADH is able to achieve chemospecific oxidation of only the secondary alcohol of a bifunctional monocyclic compound.

HLADH is capable of oxidizing a broad range of symmetric mesodiols to give the corresponding lactone products of 100% enantiomeric purity (see Fig. 6.9). Several of these lactones are excellent intermediates for important products, such as grandisol (the boll-weavil sex pheromone), pyrethroid insecticides, macrolide, and polyether antibiotics and prostaglandins.

It is unlikely that one enzyme can accomplish all the synthetic demands placed on it. This is especially true if stereospecificity of reactions is required, since this attribute usually requires a rather precise substrate–enzyme interaction. However, it is not always necessary to identify a new enzyme for each new substrate. A few enzymes of overlapping specificities are better suited to synthetic needs. One such situation is depicted in Figure 6.10, where only three alcohol dehydrogenases can collectively accommodate an enormous range of substrate structures. Furthermore, very subtle control of the stereochemistry of a product is possible using different enzymes, as illustrated in Figure 6.11.

5.4. Oxygenases

The use of oxygenases to incorporate molecular oxygen in a variety of organic substrates is being investigated in several laboratories. These en-

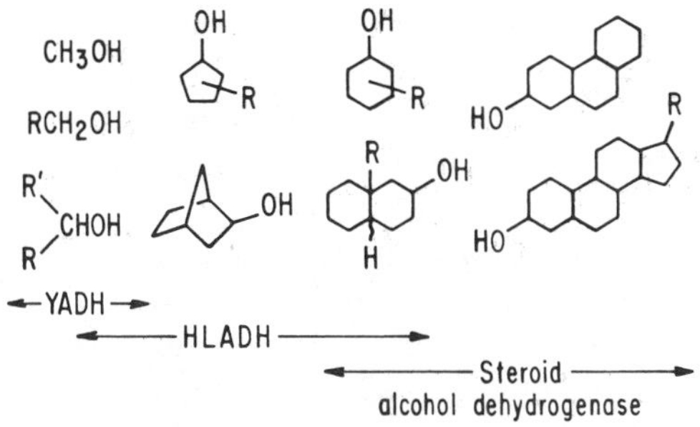

Figure 6.10. By using enzymes of overlapping substrate specifications, only three alcohol dehydrogenases are needed to access a broad range of substrate structures. YADH, yeast alcohol dehydrogenase. HLADH, horse liver alcohol dehydrogenase (Jones, 1982. Redrawn with permission.)

zymes are divided into two classes: Monooxygenases incorporate one atom of dioxygen into the substrate, the other being reduced to water at the expense of a reductant such as NADH; dioxygenases incorporate both atoms of oxygen into the substrate. These enzymes are capable of converting alkenes to alcohols, olefins to epoxides, sulfides to sulfoxides, cleaving aromatic rings, oxidatively demethylating *O*- or *N*-methyl groups, and hydroxylating aromatic and polycyclic hydrocarbons and a variety of steroid derivatives (May, 1984; Gibson, 1982). Like the reductases, the oxygenases frequently exhibit a fairly broad substrate range. For example, the enzyme methane mono-oxygenase isolated from *Methylococcus capsulatus* is capable of oxidizing ethylene, ethane, propylene, cyclohexane, benzene, toluene, methanol, styrene, and pyridine.

5.5. Problems

The reductases and oxygenases exemplify one of the problems of working with enzymes. Both classes of enzymes frequently have a requirement for an expensive and continuous source of reducing power. In the living organism

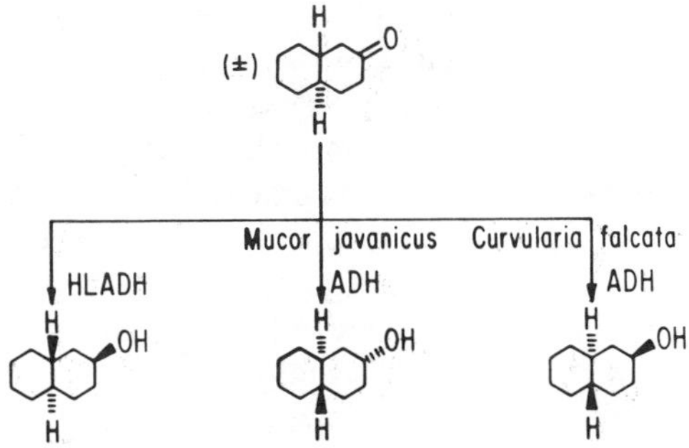

Figure 6.11. Three different 2-hydroxydecalin stereoisomers can be obtained from the same (+)-trans-2-decalone substrate using enzymes of complementary stereospecificity. (Jones, 1982. Redrawn with permission.)

this is usually provided by nicotinamide adenine dinucleotide phosphate (NADP) or by a flavin coenzyme. Techniques for the stabilization, recycling, or regeneration of these cofactors are being developed in order to use cofactor requiring enzymes for commercial synthesis (Whiteside and Wong, 1983).

Another problem with many enzymes is that they are inherently labile. Enzymes usually have a half-life measured in hours or days, whereas for commercial chemical catalysts it is usually measured in weeks or months. As has already been pointed out, genetic engineering offers considerable hope that many enzymes might be made significantly more stable to temperature, pH, organic media, and chemical and proteolytic degradation. The possible use of enzymes derived from halophilic (salt-tolerant) and thermophilic organisms, including the extremely thermophilic bacteria (growth optima from 90 to 110°C) is attracting considerable attention in this regard. For example a novel thermostable NADP-linked alcohol–aldehyde/ketone oxidoreductase has been shown to accept a broad range of substrates and to be quite stable at 85°C (Lamed and Zeikus, 1981). Also, certain marine organisms contain a unique peroxidase enzyme capable of forming heterogeneous dihalide compounds from unsaturated precursors (Neidkman and Geigert, 1983).

6. DESIGN OF SYNTHETIC PEPTIDE ENZYMES

6.1. Introduction

It is currently feasible to construct and purify analytical quantities of a peptide chain of up to 30–40 residues using essentially an updated Merrifield procedure. Recently, a Japanese group accomplished the total synthesis of a functionally active ribonuclease A, an enzyme of 124 amino acids. But this is considered a tour de force (Yajima and Fujii, 1981).

What can be done with these short peptides? A major objective is to achieve an understanding of the structural determinants of biologically active peptides to permit the design, in a rational fashion, of new peptides with comparable or enhanced activities.

Much progress has been made in the ability to predict tertiary structure from primary amino acid sequence information, but the point at which a macromolecule, such as an enzyme, possessing a great deal of tertiary structure can be designed from first principles has not yet been reached. However, progress has been made in the design and construction of active peptides in which the secondary structure rather than the tertiary structure is the dominant factor determining the function of the peptides.

Generally, secondary structure can only be assumed by peptides larger than a certain minimal size. Proteins usually have to be at least 50–60 amino acids long to achieve structural organization. Below this limit, additional structural restraints are required to ensure the maintenance of a unique structure. Thus, some small proteins achieve rigidity by the presence of numerous interchain disulfide bonds, and cyclic oligopeptides owe their unique conformation to the small peptide ring. Yet many peptides serve as biological agents of exquisite specificity even though they exist in solution in a multitude of ill-defined conformer states. In order to be able to express their specific function, these peptides must be induced to assume a special conformation. This induction is usually a result of stereospecific interactions between the ligand and the protein. It is likely that a number of peptide hormones containing less than 10 amino acids achieve their active conformation in this way.

6.2. Melittin

The main toxic compound from bee venom, melittin, is a peptide of 26 amino acids. An analysis of the amino acid sequence combined with various previous studies led Kaiser and his co-workers to the conclusion that the NH_2-terminal 20 amino acids of the peptide chain of melittin could form an amphiphilic α-helix with one side rich in hydrophobic residues and the other side relatively hydrophilic (DeGrado et al., 1981). In addition to the α-helix, there appears to be a hexapeptide active site region at the COOH terminus containing a cluster of positive charges. Melittin without this hexapeptide portion does not lyse erythrocytes but, apparently, is quite capable of binding to them.

Kaiser and co-workers designed a peptide in which the first 20 amino acids of the NH_2-terminal end of melittin were made as nonhomologous as possible to the native sequence but with as much conservation as possible of the hydrophobic or hydrophilic character of the amino acids. The complete amino acid sequence of melittin and the artificially constructed model is shown in Figure 6.12a. The hydrophobic side of the helix in the novel peptide was composed of leucine residues (6.12b) which have a high α-helix-forming potential, hydrophobicity, and electrical neutrality. Although glutamine seemed to be the optimal choice for the neutral hydrophilic residues, some serine residues were included to increase the hydrophilicity of the model, permitting the amphiphilicity of the native peptide to be matched. The α-helix breaker proline at position 14 was replaced by the α-helix-forming serine because the presence of proline was not considered important for lytic activity. Tryptophan was retained at position 19 for studies of intrinsic fluorescence, and the COOH-terminal hexapeptide was also maintained to preserve lytic activity.

Both the artificial melittin model and the native melittin form stable monolayers at the air–water interface. The artificial melittin has a higher surface affinity than melittin, consistent with a more extended helical portion for the model peptide. The hemolytic activity of the model was found to be greater than melittin, perhaps due to its enhanced surface activity. Both melittin and the model peptide bind rapidly to the outside of the erythrocyte membrane, and the surface-bound peptides produce transient openings through which

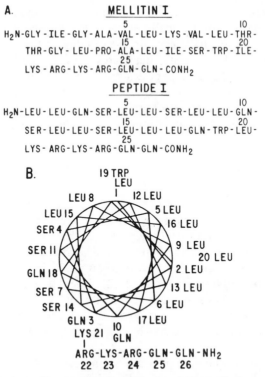

A.   MELLITIN I

5   10
H₂N-GLY-ILE-GLY-ALA-VAL-LEU-LYS-VAL-LEU-THR-
15   20
THR-GLY-LEU-PRO-ALA-LEU-ILE-SER-TRP-ILE-
25
LYS-ARG-LYS-ARG-GLN-GLN-CONH₂

PEPTIDE I

5   10
H₂N-LEU-LEU-GLN-SER-LEU-LEU-SER-LEU-LEU-GLN-
15   20
SER-LEU-LEU-SER-LEU-LEU-LEU-GLN-TRP-LEU-
25
LYS-ARG-LYS-ARG-GLN-GLN-CONH₂

B.

Figure 6.12. (*a*) Amino acid sequence of melittin I and the synthetic peptide I. (*b*) Axial projection of the α-helical region of peptide I showing the relative location of the side chains with the segregation of the hydrophobic and hydrophilic residues. (DeGrado et al., 1981. Redrawn with permission.)

hemoglobin molecules can escape. Melittin loses its ability to cause rapid lysis, presumably by being translocated through the lipid bilayer. In a substantially slower process, internalized melittin can produce transient membrane openings in the steady state. Comparison between the peptide analogue and melittin show that, on a molar basis, the synthetic analogue is similar to melittin in the initial fast process, but is actually more efficient than melittin in the secondary slow phase.

In addition to this example, several other model peptides have been constructed. For example, amphiphilic secondary structures that mimic the activity of the major human lipoproten apolipoprotein A-I (Yokoyama et al., 1980), the human hormone β-endorphin (Taylor et al., 1981), human-melanocyte-stimulating hormone α-melanotropin (Sawyer et al., 1982), and the human hormone calcitonin (Moe et al., 1983) have been synthesized.

6.3. DNA-Binding Protein

An interesting example of the *de novo* design and synthesis of a short peptide with enzymatic activity is the construction of a 34-residue peptide that exhibits strong binding to single-stranded DNA and ribonucleic acid (RNA). It has considerable ribonuclease activity with a strong preference for cleavage at the 3'-end of cytosine (Gutte et al., 1979). The peptide was designed to incorporate a succession of secondary structures, a β-strand, a reverse turn, an α-helix, another reverse turn, and an antiparallel β-strand, found in other nucleotide-binding proteins such as NAD dehydrogenase and the *E. coli* Lac repressor, although not necessarily in the same order. In addition to choosing amino acids that would fulfill the secondary structure requirements, Gutte et al. (1979) incorporated amino acids with a potential to form a number of specific noncovalent bonds with the model ligand (the trinucleotide GAA). Two additional requirements were also met. First, the binding site contained predominantly hydrophobic residues that can stack between the nucleotides and also some residues with side chains that could act as hydrogen bond donors or acceptors. Second, binding was to be achieved mainly by specific hydrogen bonds and van der Waals contacts. The interaction of the proposed structure with the trinucleotide appeared to be stabilized by including two cysteines that formed a disulfide bridge. Gutte et al. (1979) speculate that the ribonuclease activity of the model peptide may result from the careful location of a histidine residue because this residue is often found in the catalytic site of hydrolytic enzymes.

6.4. Conclusions

These examples of model peptides illustrate the power and limitations of this approach. They exemplify the construction of short secondary structure elements using the various available algorithms to predict secondary structure from primary amino acid sequence and the fact that these short peptides can exhibit biological activity. On the other hand, it is not feasible to synthesize peptides much longer than 30–40 residues, nor is it possible to predict tertiary structure from a linear amino acid sequence. Therefore, the de novo synthesis of even modest enzymes (about 100 amino acid residues) is not currently possible. The real advantage to be gained from the construction and study of these model peptides is an improved understanding of the relationship between the structure and function of enzymes and other proteins.

7. NONPEPTIDE ENZYME MIMICS

7.1. Introduction

Enzymes have several potential disadvantages as industrial catalysts. They generally require water for activity, although as we have seen this need not necessarily be true for all enzymes. Many enzymes are unstable under certain conditions of temperature, pressure, and pH or can be degraded by proteolytic enzymes or by chemical oxidation. In addition, many enzymes require cofactors for activity. Although cofactors are used only in catalytic amounts, they are generally expensive, and only in certain cases can they be inexpensively regenerated after catalysis. Synthetic catalysts are currently being designed in several laboratories that may exhibit the key advantages of enzymes, namely, specificity of substrate, specificity of reaction, and rate enhancement. It is hoped that these nonpeptide based enzyme mimics will not suffer the potential drawbacks of naturally occurring enzyme systems.

Since most enzymes incorporate a concave surface or cavity for substrate binding, researchers have focused on designing organic molecules with binding cavities, called cavitands, followed by attachment of appropriate catalytic sites to these cavitands.

7.2. Cyclodextrins

Several years ago, Breslow and his colleagues built a catalyst based on cyclodextrin (Breslow et al., 1976; Breslow and Campbell, 1969). Cyclodextrin is a doughnut-shaped molecule made up of glucose units with an interior whose size and shape are determined by the number of glucose units that make up the ring. Cyclodextrin is soluble in water by virtue of the hydroxyl groups of the glucose that rim the cavity. However, the interior of the molecule is relatively hydrophobic and is able to withdraw small organic molecules from the aqueous surroundings into this cavity. This is similar to the ability of an enzyme to bind its substrate in its interior cavity. Furthermore, like an enzyme, the cyclodextrin binding is selective for molecules with the correct shape and hydrophobic character.

As shown in Figure 6.13a a cyclodextrin was constructed that would permit the attachment of a chlorine atom to only one specific position on an anisole molecule. In the absence of cyclodextrin, anisole is randomly chlorinated to produce a mixture of orthochloroanisole and parachloroanisole. However, because cyclodextrin binds anisole in its cavity it prevents the chlorination of the anisole in the ortho position, but permits parachloroanisole to be produced by a new chemical pathway, catalyzed by a hydroxyl group of the cyclodextrin.

This chlorination system exhibits many enzyme-like properties. The cyclodextrin selectively binds particular molecules and then selectively produces a single product by a catalyzed reaction within the complex. Finally, the product, parachloroanisole, is released and a new anisole substrate molecule binds to start the cycle again. Further work demonstrated that it was possible to extend the principle of the cyclodextrin-based catalyst to include other substrates and other reaction specificities (Breslow, 1982).

However, in all the earlier studies and other related ones, only modest enhancements of reaction rates were achieved. One principle, known for some time, is that an enzyme can bind the transition state of a reacting complex better than it can bind the initial substrate. This activity is a major

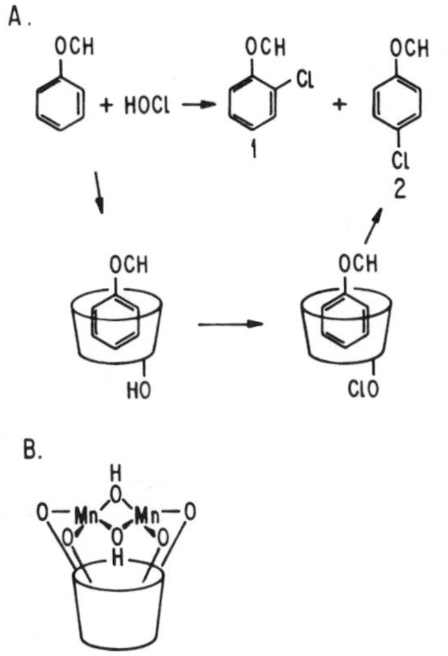

A.

OCH

$\bigcirc$ + HOCl $\longrightarrow$ OCH
Cl
1
+ OCH
Cl
2

OCH
HO
$\longrightarrow$
OCH
ClO

Figure 6.13. (*a*) The random chlorination of anisole in solution is converted to a selective process by a cyclodextrin catalyst, which binds the anisole in a cavity and delivers chlorine under geometric control (Breslow et al., 1976. Redrawn with permission.) (*b*) A Mn-coupled β-cyclodextrin used to model the activity of the water-splitting enzyme in photosynthesis. (Nair and Dismukes, 1983.)

B.

factor in determining the enhanced reaction rate of an enzymatic process. But it was found that the cyclodextrins were binding the substrates better than their respective transition states. Consequently, a cyclodextrin was modified by the addition of intrinsic groups that produced a floor in the cavity that reduced the binding efficiency of the substrate. This resulted in an increase of about 10-fold in reaction rate because the new geometry presumably favors the transition state (Breslow, 1982).

Additional rate enhancements could be achieved by the use of molecules based on the ferrocene nucleus as a substrate (Breslow et al., 1983). Ferrocene, a compound consisting of an iron atom sandwiched between two cyclopentadiene rings, fits a cyclodextrin extremely well. If a *p*-nitrophenyl ester of ferroceneacrylic acid is used, a 10^5 enhancement of hydrolysis is observed because the flexible side chain brings the ester near to the catalytic site. If the acrylic side chain is incorporated into a fused ring system via a double bond so that the ester is held near the hydroxyl more rigidly (see Fig. 6.14), a rate enhancement of about 6×10^6 can be obtained compared to the uncatalyzed reaction. Furthermore, the reaction is carried out in aqueous dimethyl sulfoxide in order to solubilize the reactants. The uncatalyzed reaction in dimethyl sulfoxide proceeds 24 times faster than the reaction in

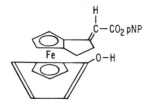

Figure 6.14. A schematic representation of modified ferrocene interacting with a cyclodextrin molecule. (Breslow, 1982. Redrawn with permission.)

water, so the overall rate increase compared to hydrolysis in water is about 1.5×10^8, well within the range of accelerations accomplished by enzymes. In addition, in one instance, the optical activity of the cyclodextrin nucleus led to a selection of one of two mirror-image isomers of the substrate in the ratio of 65:1.

Another example of a cyclodextrin enzyme-mimic is a Mn-coupled β-cyclodextrin used to model the activity of the binuclear clusters of Mn in the enzyme that catalyzes the oxidation of water to oxygen during photosynthesis (Nair and Dismukes, 1983). Two atoms of Mn (III) were attached to the cyclodextrin through an oxolinkage at the top of the molecule (Fig. 6.13b). This may mimic the attachment of the Mn to the enzyme, which probably involves an oxygen ligand. Furthermore, the complex exhibits reversible electrochemistry. However, its ability to split water remains to be demonstrated. The importance of this study is the demonstration of the feasibility of modeling the metal clusters used in enzymatic processes. As we pointed out earlier, enzymes that carry out oxidations usually require a metal or other cofactor because no amino acid chains are good electron acceptors.

There are many other examples of experiments to model metal and other ligand interactions with the enzymes. Recently, Lippard and co-workers constructed an accurate model of the di-iron center of hemerythryn, an enzyme capable of reversibly binding and transporting oxygen (Armstrong et al., 1984). The model may also mimic the activity of the active center of ribonucleotide reductase, the catalyst that converts ribonucleotides to deoxyribonucleotides.

7.3. Chymotrypsin Mimic

Cram and his co-workers have synthesized a model catalyst that mimics the activity of the enzyme chymotrypsin (Cram and Katz, 1983). As we described earlier, chymotrypsin enhances the rate of hydrolysis of peptide bonds by about 10^8–10^{10} compared to the rate in the absence of the enzyme. This catalytic activity requires a fairly complex structure that is designed to recognize and bind the substrate. It can also direct the appropriate part of the substrate to the active or catalytic site and can respond in a dynamic way to the substrate binding resulting in molecular motion of the critical amino acids in the catalytic site. This movement establishes a charge relay system that activates the normally unreactive hydroxyl group of a precisely positioned serine residue and so initiates the hydrolytic cleavage. Clearly, the enzyme is structured not only to have a suitably located substrate binding and catalytic site but also to have a protein framework that will permit molecular motion of critical components. On the other hand, the Cram group has built a relatively simple "enzyme," or host, consisting of a cage structure of modified anisole and cyclic urea units. These are designed with suitable appendages, such as oxygen atoms, to form a cavity that will bind

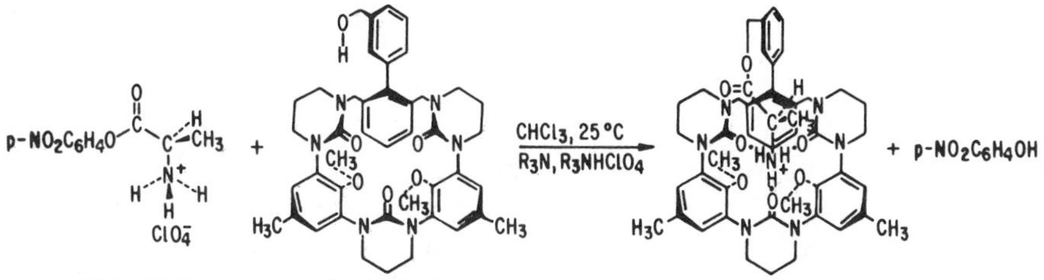

Figure 6.15. Model for the acylation of an alanine ester using a synthetic cage structure to enhance the reaction rate. (Cram and Katz, 1983. Redrawn with permission.)

the substrate comfortably, which in this case, is the protonated L-alanyl *p*-nitrophenyl ester cation (see Figure 6.15). Binding brings a strategically positioned hydroxyl nucleophile on the host close to the guest, which rapidly forms an acyl bond across the oxygen of the hydroxyl group. The rate enhancement of this step is accelerated 10^{10} times compared to a similar reaction in which the guest is mixed with a nucleophile lacking the ability to form a complex with the ester of alanine. As yet the final steps, hydrolysis and decarboxylation of the guest, are not carried out by the host, so strictly speaking the host is not a catalyst. The next step will be to design a host that will bring in the water molecule and promote removal of the reaction products. This is accomplished in chymotrypsin by an imidazole ring of a histidine and the carboxylate of an aspartate residue. Thus an imidazole and a carboxylate group will be added by Cram to the model host in order to achieve hydrolysis and turnover.

8. THE FUTURE

In the future, our knowledge of the versatility of enzyme reactions will be greatly extended. This will arise both from an enhanced familiarity with enzymes by organic chemists and biochemists and because a greater number of enzymes will become available through genetic engineering. There will be continued research into the modification of substrate specificities, reaction specificities, and reaction mechanisms in organic media and at different temperatures and pH's. There will be a marked interest in enzymes derived from various exotic microorganisms such as thermophiles and halophiles with the objective of obtaining new enzymes of industrial and scientific importance.

Our knowledge of the relationship between enzyme structure and function will be enhanced. This will be due to (1) the availability of more models of enzymes at the atomic level. The rate at which information is becoming available is increasing because of the greater availability of purified enzymes, faster techniques for protein crystallization, the availability of protein sequences by rapid DNA sequencing techniques, the rapid collection of data by recently developed synchrotron x-ray techniques, and the availability of new information on protein dynamics derived from nuclear magnetic resonance (NMR) studies. (2) Improved methods for data analysis especially interactive computer graphics for visualizing protein structures and interactions. (3) New data from the modeling of short peptides will greatly improve the algorithms for predicting protein structure and function from the linear amino acid sequence. Also data from the use of nonpeptide enzyme mimics will help elucidate the requirements for substrate specificity and reaction type. (4) Site-directed mutagenesis and other genetic engineering techniques will similarly revolutionize our understanding of enzyme structure and function.

There will also be a requirement for biochemical engineers to implement the new technology and develop further techniques for enzyme use. Above all, the development of the enzyme business will require a new synergism among the disciplines of chemistry, biochemistry, and cell and molecular biology. The biochemical engineer of the future should have a firm grounding in these disciplines as well as in engineering. To accomplish this, a new effort should be made by the universities to provide appropriate multidisciplinary courses at the graduate level. Our approach to the training of biochemical engineers lags severely behind the technical advances in the basic sciences that contribute to the discipline of biochemical engineering.

REFERENCES

Armstrong, W. H., Spool A., Papaefthymiou, G. C., Frankel, R. B., and Lippard S. J. Assembly and characterization of an accurate model for the diiron center in hemerythin. *J. Am. Chem. Soc.* **106**, 3653–3667, 1984.

Breslow, J. Artificial enzymes. *Science* **218**, 532–536, 1982.

Breslow, R., and Campbell, P. Selective aromatic substitution with a cyclodextrin mixed complex, *J. Am. Chem. Soc.* **91**, 3085–3087, 1969.

Breslow, R., Kohn, H., and Siegel, B. Methylated cyclodextrin and a cyclodextrin polymer as catalysts in selective anisole chlorination, *Tetrahedron Lett.* **20**, 1645–1646, 1976.

Breslow, R., Trainor, G., and Ueno, A. Optimization of metallocene substrates for β-cyclodextrin reactions. *J. Am. Chem. Soc.* **105**, 2739–2744, 1983.

Cambou, B., and Klibanov, A. M. Preparative production of optically active esters and alcohols using esterase-catalyzed stereospecific transesterification in organic media. *J. Am. Chem. Soc.* **106**, 2687–2692, 1984.

Cram, D. J., and Katz, H. E. An incremental approach to hosts that mimic serine proteases. *J. Am. Chem. Soc.* **105**, 135–137, 1983.

DeGrado, W. F., Kezdy, F. J., and Kaiser, E. T. Design, synthesis and characterization of a cytotoxic peptide with melittin-like activity. *J. Am. Chem. Soc.* **103**, 679–681, 1981.

Estell, D. A., Miller, J. V., Graycar, T. P., Powers, D. B., and Wells, J. A. Site-Directed Mutagenesis of the Active Site of Subtilisin BPN'. Biotech '84 USA, 1984, Pinnar, UK, Online Publ., pp. 181–187, 1984.

Gibson, D. T. Microbial degradation of hydrocarbons, *Toxicological Environ. Chem.* **5**, 237–250, 1982.

Gutte, B., Daumigen, M., and Wittschieber, E. Design, synthesis and characterization of a 34-residue polypeptide that interacts with nucleic acids, *Nature* **281**, 650–655, 1979.

Inada, Y., Nishimura, H., Takahashi, K., Yoshimoto, T., Saha, A. J., and Saito, Y. Ester synthesis catalyzed by polyethylene glycol-modified lipase in benzene, *Biochem. Biophys. Res. Comm.* **122**, 845–850, 1984.

Jay, E., MacKnight, D., Lutze-Wallace, C., Harrison, D., Wishart, P., Liu, W-Y., Asundi, V., Pomeroy-Cloney, L., Rommens, J., Eglington, L., Pawlak, J., and Jay, F. Chemical synthesis of a biologically active gene for human immune interferon-γ. *J. Biol. Chem.* **259**, 6311–6317, 1984.

Johanson, C. Lecture at the Biotech 1984 USA Conference in Washington, D.C. (not published in the proceedings), 1984.

Jones, J. B. Horse liver alcohol dehydrogenase: An illustrative example of the potential use of enzymes in organic synthesis. In Chibata I., Fukui S., and Wingard, L. B. (eds.). New York, Plenum Press, 1982, pp. 107–116.

Kaiser, E. T., Levine, H. L., Otsuki, T., Fried, H. E. and Dupeyre, R-M. Studies on the mechanism of action and stereochemical behavior of semisynthetic model enzymes. In D. Dolphin, et al. (eds.). *Biomimetic Chemistry. Advances in Chemistry* Series 191. Washington, D.C., Am. Chem. Soc., 1980, pp. 35–48.

Klibanov, A. M. Immobilized enzymes and cells as practical catalysts. *Science* **219**, 722–727, 1983.

Lamed, R. J., and Zeikus J. G. Novel NADP-linked alcohol-aldehyde/ketone oxidoreductase in thermophilic ethanologenic bacteria. *Biochem. J.* **195**, 183–190, 1981.

Margalit, R., Pecht, I., and Gray, H. Oxidation-reduction catalytic activity of a pentaamineruthenium (III) derivative of sperm whale myoglobin. *J. Am. Chem. Soc.* **105**, 301–302, 1983.

Maugh T. H. A renewed interest in immobilized enzymes. *Science* **223**, 474–475, 1984.

May, S. Synthetic transformation with oxidoreductase enzymes. Biotech '84 USA. Online Publ. UK, Pinner, 1984, pp. 199–204.

Moe, G. R., Miller, R. J., and Kaiser, E. T. Design of a peptide hormone: Synthesis and characterization of a model peptide with calcitonin-like activity. *J. Am. Chem. Soc.* **105**, 4100–4102, 1983.

Nabiar, K. P., Stackhouse, J., Stauffer, D. M., Kennedy, W. P., Elridge, S. K., and Benner, S. A. Total synthesis and cloning of a gene coding for the ribonucleases protein. *Science* **223**, 1299–1301, 1984.

Nair, B. V., and Dismukes, G. C. Models for the photosynthetic water oxidizing enzyme. I. A binuclear manganese (III)-β-cyclodextrin complex. *J. Am. Chem. Soc.* **105**, 124–125, 1983.

Nakanishi, K., Kamikub, T., and Matsumo, R. Continuous synthesis of N-(benzyloxycarbonyl)-L-aspartyl-L-phenylalanine methyl ester with immobilized thermolysin in an organic solvent. *Biotechnology* **3**, 459–464, 1985.

Neidleman, S. L., and Geigert, J. The enzymatic synthesis of heterogeneous dihalide derivatives: A unique biocatalytic discovery. *Trends Biotechnol.* **1**, 21–25, 1983.

Orr, T. Non-crystallographic short-cuts allowing protein engineering to progress to enzymes. *Gen. Eng. News,* pp. 20–21, Oct. 1984.

Sawyer, T. K., Hruby, J., Darman, P. S., and Hadley, M. E. [half-Cys[4], half-Cys[10]]-α-melanocyte-stimulating hormone: A cyclic α-melanotropin exhibiting superagonistic biological activity. *Proc. Nat. Acad. Sci. USA* **79**, 1751–1755, 1982.

Slama, J. T., Oruganti, S. R., and Kaiser, E. T. Semisynthetic enzymes: Synthesis of new flavopapain with high catalytic efficiency. *J. Am. Chem. Soc.* **103**, 6211–6213, 1981.

Stryer, L. *Biochemistry,* 2nd ed. New York, Freeman, 1981.

Taylor, J. W., Miller, R. J., and Kaiser, E. T. Design and synthesis of a model peptide with β-endorphin-like properties. *J. Am. Chem. Soc.* **103,** 6965–6966, 1981.

Tuengler, P., and Pfleiderer, G. Enhanced heat, alkaline and tryptic stability of acetaminated pig heart lactate dehydrogenase. *Biochim. Biophys. Acta* **484,** 1–8, 1977.

Ulmer, K. Protein engineering. *Science* **219,** 666–671, 1983.

Whitesides, G. M., and Wong, C-H. Enzymes as catalysts in organic synthesis. *Aldrichemica Acta* **16,** 27–34, 1983.

Wilkinson, A. J., Fersht, A. R., Blow, D. M., Carter, P., and Winter, G. A large increase in enzyme-substrate affinity by protein engineering. *Nature* **307,** 187–188, 1984.

Yajima, H., and Fujii, N. Studies in peptides. 103. Chemical synthesis of a crystalline protein with the full enzymatic activity of ribonuclease A. *J. Chem. Soc.* **103,** 5867–5871, 1981.

Yokoyama, S., Fukushima, D., Kezdy, F. J., and Kaiser, E. T. The mechanism of activation of lecithin: Cholesterol acyltransferase by apolipoprotein A-I and an amphiphilic peptide. *J. Biol. Chem.* **255,** 7333–7339, 1980.

Zaks, A., and Klibanov, A. M. Enzyme catalysis in organic media at 100°C. *Science* **224,** 1249–1251, 1984.

Zubay, G. *Biochemistry,* Reading, MA, Addison-Wesley, 1983.

ADDITIONAL READING

General

Craik, C. S. Use of oligonucleotides for site-specific mutagenesis. *Biotechniques* Jan/Feb. 12–19, 1985.

Drexler, K. E. An approach to the development of general capabilities for molecular manipulation. *Proc. Nat. Acad. Sci.* USA **78,** 5275–5278, 1981.

Fox, J. L. Protein engineering. *Chemical and Engineering News* **51,** 566–568, 1985.

Fox, J. L. Synthetic catalysts approach enzyme efficiency. *Chemical and Engineering News* Feb. 14, 1983, 33–34.

Katchalski-Katzir, E. Recent developments and future aspects of enzyme engineering. In Chibata, I, et al. (ed.) *Enzyme Engineering.* New York, Plenum Press, 1983, pp. 503–510.

Katchalski-Katzir, E., and Freeman, A. Enzyme engineering reaching maturity. *Trends Biol. Sci.* 427–431, Dec. 1982.

Layman, P. L. Enzyme business attracting new producers, technology, *Chemical and Engineering News,* Sept. 12, 1983, 11–15.

Maugh, T. Semisynthetic enzymes are new catalysts. *Science* **223,** 154–156, 1984.

Maugh, T. Need a catalyst? Design an enzyme. *Science* **223,** 269–271, 1984.

Maugh, T. Catalysts that break nature's monopoly. *Science* **221,** 351–354, 1983.

Drexler, K. E. Molecular engineering: An approach to the development of general capabilities for molecular manipulation. *Proc. Natl. Acad. Sci. USA* **78,** 5275–5278, 1981.

Pabo, C. Designing proteins and peptides. *Nature* **301,** 200, 1983.

Researchers close in on synthetic enzyme. *Chemical and Engineering News,* April 1979, 26–27.

Stewart, M. The role of the biochemical engineer. *Biotech Europe* **84,** 247–264 (Online Conferences Ltd, UK), 1984.

Chemical Modification of Proteins

Kaiser, E. T., Lawrence D.S., and Rokita S.E. The chemical modification of enzymatic specificity. *Ann. Rev. Biochem.* **54,** 565–595, 1985.

Modification of Proteins by Site-directed Mutagenesis

Dodge, A. R. Stability of enzymes from thermophilic microorganisms. Adv. Appl. Environ. Microbiol. **23,** 17–21, 1977.

Itakura, K., et al. Chemical synthesis and application of oligonucleotides of mixed sequence, in Walton, A. G. (ed.). *Recombinant DNA,* Proc. Third Cleveland Symp. on Macromolecules, Cleveland, Ohio. Amsterdam, Elsevier, 1981, pp. 273–289.

Rossi, J. J., et al. The role of synthetic DNA in the preparation of structural genes coding for proteins, in *From Gene to Protein: Translation into Biotechnology,* New York, Academic Press, 1982, pp. 213–234.

Smith, M. Site-directed mutagenesis. *Trends Biol. Sci.,* December, 440–492, 1982.

Villefranca, J. E., et al. Directed mutagenesis of dihydrofolate reductase. *Science* **222,** 782–788 1983.

Design of Synthetic Peptide Enzymes

Chaiken, I. M. Semisynthetic peptides and proteins. *CRC Crit. Rev. Biochem.* **11**(3), 255–301, 1981.

Kaiser, E. T., and Kezdy, F. J. Secondary structure of protein and peptides in amphiphilic environments. *Proc. Natl. Acad. Sci.* USA **80,** 1137–1143, 1983.

Offord, R. E. Semisynthetic Proteins. New York, John Wiley, 1980.

Designing Proteins, Protein Modeling

Chou, P. Y., and Fasman, G. D. Empirical predictions of protein conformation. *Ann. Rev. Biochem.* **47,** 251–276, 1978.

Connolly, M. L. Solvent-accessible surfaces of proteins and nucleic acids. *Science* **221,** 709–713, 1983.

Dayhoff, M. O., and Eck, R. V. Atlas of protein sequence and structure. (1983). *Nat. Biomed. Res. Found.,* Silver Springs, Maryland, 1983.

Heremans, K. High pressure effects on proteins and other biomolecules. *Ann. Rev. Biophys. Bioeng.* **11,** 1–21, 1982.

Jaenicke, R. Enzymes under extremes of physical conditions. *Ann. Rev. Biophys. Bioeng.* **10,** 1–67, 1981.

Karplus, M., and McGammon, J. A. Dynamics of proteins: elements and function. *Ann. Rev. Biochem.* **53,** 263–300, 1983.

Kim, P. S., and Baldwin, R. L. Specific intermediates in the folding reactions of small proteins and the mechanisms of protein folding. *Ann. Rev. Biochem.* **51,** 459–489, 1982.

Maugh, T. A new dimension in NMR. *Science* **224,** 46–47, 1984.

Orcutt, B. C., George, D. G., and Dayhoff, M. O. Protein and nucleic acid sequence database systems. *Ann. Rev. Biophys. Bioeng.* **12,** 419–441, 1983.

Tucker J. B. Designing molecules by computer. *High Technology,* Jan., 52–59, 1984.

Wolfenden, R. Waterlogged molecules. *Science* **222,** 1087–1093, 1983.

7

APPLIED GENETICS AND MOLECULAR BIOLOGY OF INDUSTRIAL MICROORGANISMS

RICHARD P. ELANDER

Vice-President, Biotechnology
Bristol-Myers Company
Industrial Division
Syracuse, New York

1. INTRODUCTION

During the past three decades, genetic manipulation of microbial pathways for antibiotic synthesis has generated a large number of clinically important agents. There are now approximately 160 antibiotic compounds available for chemotherapy. Antibiotics that become commercially important are discovered and further developed by research efforts aimed at (1) screening microorganisms for naturally occurring secondary metabolites, (2) modifying natural substances chemically to produce semisynthetic derivatives having desirable characteristics, and (3) improving strain techniques to select for mutants that can produce large amounts of useful antibiotics. Compounds must be both cost and health effective in order to become marketable. Much of the microbiological and genetic research in the pharmaceutical industry is, therefore, focused on the development of effective strategies for manipulating microorganisms to produce greater quantities of known antibiotics and novel antibiotic compounds. Approximately 6500 naturally occurring and over 32,000 semisynthetically derived antibiotics have been discovered since the early part of this century (Demain, 1983). In most cases, these compounds were obtained or identified by empirical methods (Elander et al., 1973). There have also been great improvements in the antibiotics productivity levels of particular microoganisms, which have resulted from essentially random approaches. The future of antibiotic screening and strain improvement technology will move away from the empirical, random methods of the past toward the application of more rigorous scientifically approaches. This will happen because of advances in the understanding of the antibiotic biosynthetic pathways in many organisms (Ball and Azevedo, 1977), elucidation of regulatory mechanisms related to the induction and repression of the genes involved in antibiotic synthesis (Ball and Azevedo, 1976), and knowledge of the physiology of the relevant microorganisms. It will be possible to develop and apply new nonempirical strategies for both strain improvement and novel antibiotic development. Genetic engineering techniques, including recombinant deoxyribonucleic acid (DNA) methods, can facilitate the application of some of these new strategies (Macdonald and Holt, 1976; Elander, 1981). This chapter focuses on the applied genetics of antibiotic-producing microorganisms and will center on three primary objectives:

1. Genetic improvement of microorganisms leading to the development of strains that produce large quantities of known antibiotics
2. Altering the genetic structure of antibiotic-producing microorganisms in a controlled but semirandom fashion to facilitate the biological synthesis of new, improved antibiotics
3. Strategies for the specific alteration of antibiotic synthesis pathways to create new, slightly modified strains capable of the biological production of known antibiotics currently manufactured by semisynthetic chemical methods

2. RATIONAL SELECTION

2.1. Improvement in Pyrrolnitrin Formation by Analogue-Resistant Mutations

Mutants resistant to amino acid analogues have been used successfully for the overproduction of amino acids (Adelberg, 1958; Arima, 1977). Application of a rational approach to secondary metabolites was reported by Elander et al. (1971) for pyrrolnitrin production by strains of *Pseudomonas fluorescens*. In this example, D-tryptophan was a direct precursor of pyrrolnitrin (Figure 7.1), an antifungal antibiotic, but was impractical to use in a

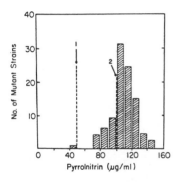

Figure 7.1. Pyrrolnitrin production by a 6-fluorotrypto-phan (6-FT)-resistant population of *P. fluorescens*. (1) Yield of drug-sensitive grandparent A10338.5; (2) yield of first-step 5-FT-resistant parent A10338.7 (From Elander et al.) 1971.

large-scale fermentation. Mutants resistant to fluoro- or methyl-tryptophan were selected and produced nearly threefold more antibiotic than the sensitive parent; in addition, D-tryptophan was no longer necessary for maximal pyrrolnitrin formation.

2.2. Improved β-Lactam Antibiotic Production

α-Aminoadipic acid, cysteine, and valine are direct biosynthetic precursors of the important β-lactam antibiotics of the penicillin and cephalosporin class. Mutants of *Penicillium* and *Cephalosporium* that were resistant to analogues of these and related amino acids were selected, and their cephalosporin C productivity was examined. In comparison with random selection, mutants resistant to certain methionine or lysine analogues yielded a higher frequency of superior strains. Although methionine is not a precursor for cephalosporin C, it was hoped that mutants resistant to methionine analogues such as trifluoromethionine or selenomethionine would overproduce methionine (either endogenously or extracellularly) thus alleviating the need for added methionine. No efforts were made to determine whether the resistant mutants actually overproduced the antagonized amino acid or were simply deficient in uptake of the toxic analogue.

Ions of heavy metals such as mercuric ions (Hg^{2+}), cupric ions (Cu^{2+}), and related organometallic ions complex with β-lactam producing organisms at high concentrations. We can theorize that many of the mutants that become resistant to these metallic ions may do so by overproducing β-lactam antibiotics as a means of detoxifying these metallic ions or their possible interference with sulfhydryl compounds either as substrate intermediates or on enzyme-active sites. Mutants have been isolated for resistance to phenyl-mercuric acetate (PMA) for improved cephamycin production in *Streptomyces lipmanii* (Godfrey, 1973). At Bristol-Myers Company, we have isolated mutants resistant to inhibitory concentrations of Cu^{2+}, PMA, chromate, Mn^{2+}, and Hg^{2+} and tested them for cephalosporin C production. The frequency of superior isolates among mutants resistant to mercuric chloride was higher than that for isolates selected randomly among survivors of ultraviolet (UV) treatment. Isolation of mutants resistant to organometallic compounds has been used for selection of specific types of nutritional mutants. A new class of methionine auxotrophs in *Saccharomyces cerevisiae* have been obtained by isolating methyl mercury-resistant mutants (Singh and Sherman, 1974).

One of the difficulties in applying the resistant mutant approach to strain improvement is that many analogues either do not inhibit growth or inhibit growth only at high concentrations. Furthermore, some analogues reduce colony size or inhibit sporulation but do not inhibit colony formation at high concentrations. In these cases, strains that form normal-size or sporulating colonies in the presence of inhibitors are considered resistant.

In addition to selecting mutants resistant to amino acid analogues, mutants resistant to various metabolic inhibitors may have altered permeability characteristics, thereby affecting the regulation and secretion of secondary metabolites. Penicillin V and cephalosporin C yield improvement programs at Brisol-Myers Company based on rational selection procedures that are summarized in Table 7.1.

2.3. Nocardicin Production in Strains of *Nocardia uniformis*

Two major problems were encountered during the initial stages of strain improvement for nocardicin A production with *N. uniformis* subsp. *tsu-*

TABLE 7.1. Comparison of Random versus Directed Selection Procedures in Strains of *Penicillium chrysogenum* and *C. acremonium*[a]

Type	Mutagenesis	Organism	Selected Method	% Retained[b]
Random	UV, NG[c]	*P. chrysogenum* and *C. acremonium*	None	0.81
Directed	UV, NG X ray	*P. chrysogenum* and *C. acremonium*	Colony plate	1.36
			Auxotrophs	5.71
			Haploidization agents	1.59
			Mitotic inhibitors	1.33
			Amino acid analogues	1.41
			Sulfur analogues	3.98
		C. acremonium	Methionine analogues	3.64
			Increased methionine sensitivity	
			Growth	1.62
			β-Lactam synthesis	3.85

[a] Adapted from Chang and Elander (1979).
[b] Superior on both primary and secondary screening tests.
[c] NG, nitrosoguanidine; UV, ultraviolet.

yamanensis (Elander and Aoki, 1982). First, this *Nocardia* strain was found to be sensitive to β-lactam antibiotics including nocardicin A, which was the product of the strain itself. When nocardicin (2 mg/ml) was added to strain No. 1923, its growth was inhibited. The cells were fragmented and lysed. A strain improvement program was initiated for the selection of mutants that were able to form colonies on agar plates containing varying concentrations of nocardicin A. Mutagenic treatment with nitrosoguanidine or UV radiation produced mutants resistant to nocardicin A in varying degrees.

Another problem was β-lactamase production, which destroyed nocardicin A and other β-lactam antibiotics upon formation. The β-lactamase was secreted into the culture medium. Among nocardicin-A-resistant mutants, strains with increased β-lactamase activity were frequently observed. Thus, the next step in strain improvement programs was to select β-lactamase negative mutants from among the nocardicin-A-resistant strains. A simple method was devised for the detection of β-lactamase production. Mutagen-treated spore populations were plated on an agar medium. Soft agar, containing a dilution of a culture of *Staphylococcus aureus* 209 P and penicillin G (10 μg/ml) was overlayed and incubated at 37°C for 18 hr. A growth zone of *Staphylococcus* cells appeared around the colonies with β-lactamase that decomposed penicillin G. β-lactamase-negative mutants were selected as colonies that did not show detectable growth zones around them. Stable isolates were selected from the β-lactamase-negative mutants, which had not reverted in β-lactamase production in subsequent strain-improvement efforts. Once the productivity of β-lactamase was lost, the degree of resistance to nocardicin A proved to be a practical technique for the selection of superior mutants (Figure 7.2). The frequency of appearance of variants superior to the parent in nocardicin A productivity was higher when the mutagen-treated spores were plated and colonies were selected from a medium containing nocardicin A at a concentration that inhibited the growth of the parent strain.

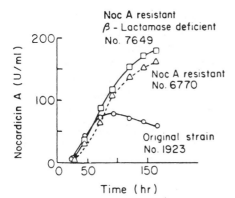

Figure 7.2. Improved nocardicin A production by β-lactamase-deficient/nocardicin-A-resistant mutants of *N. uniformis*. (From Aoki, personal communication.)

2.4. Rational Selection/Protoplast Fusion Strategies for Improved Carbapenem Production

Streptomyces griseus subsp. *cryophilus* produces the novel carbapenem antibiotics, C-19393 H_2 and S_2 (carpetimycin A and B[2]) and E_5, together with epithienamycins and olivanic acids. These compounds are classified into two categories that is, sulfated (C-19393 S_2, MM-4550, MM-13902, and MM-17880) and unsulfated (C-19393 H_2, E_5, epithienamycins A, B, C, and D) carbapenem antibiotics. In a chemically defined medium, the production ratio of these two groups of the antibiotics was affected by the sulfate concentration in the medium. Mutants unable to produce sulfated carbapenem antibiotics were successfully derived as sulfate transport-negative mutants (Kitano et al., 1982). Auxotrophs requiring thiosulfate or cysteine for growth or mutants resistant to selenate produced twofold more unsulfated carbapenem antibiotics as did the parental strain C-19393. Converged accumulation of C-19292 H_2 was achieved by selecting strains resistant to S-2-aminoethyl-L-cysteine from a sulfate transport-negative mutant, K-4. A mutant, K4-38-LHx-1, produced C-19393 H_2 as the major component and produced 10-fold more antibiotic when compared to the original strain. The sulfate transport system was reintroduced into sulfate transport-negative mutants, including K4-38-LHx-1 from the strain C-19393 through protoplast fusion. Stable genetic recombinants were obtained at high frequency by treating the mixed protoplasts with 40% polyethylene glycol 4000 and high producers of C-19393 S_2 were found among these recombinants.

2.5. Cerulenin-Resistant Mutants for Improved Daunorubicin Production

Rational selection procedures for strain improvement of daunorubicin, an anthracycline antitumor antibiotic (McGuire et al., 1980), were based on the polyketide origin of daunorubicin and on the ability of the antibiotic cerulenin to inhibit polyketide biosynthesis. Cerulenin, at a concentration of about 50 μM, suppresses the formation of anthracycline pigment. Occasional pigmented colonies stood out against a background of unpigmented colonies on cerulenin agar. The pigmented colonies were retested for pigment production on cerulenin agar and were then tested for production of daunorubicin glycosides. Figure 7.3 shows a histogram of the daunorubicin glycoside titers produced by a series of mutants isolated by the cerulenin screen. Mutants were equally divided into three groups—those producing the same amount of daunorubicin glycoside as the starting strain, those producing more, and those producing substantially less. The very low producers (< 10 μg/ml) were found to accumulate high levels of the anthracycline aglycone ϵ-rhodomycinone, hence their formation of red colonies on cerulenin agar.

Many other important antibiotics are biosynthesized by the polyketide

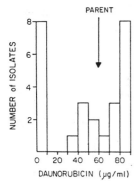

PARENT

Figure 7.3. Daunorubicin production by cerulenin-resistant mutants. (From McGuire et al., 1980.)

pathway. Examples include the tetracyclines, rifamycins, erythromycins, and anthracyclines. It would be interesting to apply cerulenin resistance to organisms producing these polyketide-derived metabolites as a rational means to isolate improved mutants.

2.6. Macrolide, Lincosamide, and Streptogramin Resistance and Improved Macrolide Antibiotic Formation

Inducible resistance to macrolide, lincosamide, and streptogramin (MLS) type B antibiotics occurs in a wide range of *Streptomyces* sp. and comprises a family of related response patterns in which subsets of MLS antibiotics characteristic of a particular strain induce resistance to all MLS antibiotics (Fujisawa and Weisblum, 1981). The induced phenotype is a consequence of specific N6-methylation of adenine in 23S ribosomal ribonucleic acid (rRNA). As a result, the ribosome becomes coresistant to all MLS antibiotics. *Streptomyces lincolnensis* NRRL 2936 (lincomycin producer) lacks methylated adenine in 23S rRNA during active growth. Supplementation of the growth medium with lincomycin (1 μg/ml) resulted in the appearance of methylated adenine in 23S or rRNA. A constitutively resistant mutant of *S. lincolnensis* selected with maridomycin and grown in the absence of antibiotics was found to contain from one to two residues of N6-monomethyladenine in 23S rRNA. Lincomycin production and appearance of methylated rRNA may be functionally linked; cells cannot undertake lincomycin synthesis without first becoming lincomycin resistant. Weisblum (1982) tested the hypothesis that constitutively resistant cells of *S. lincolnensis* might produce lincomycin more efficiently than wild-type cells that require induction. This approach was extended to a tylosin-producing strain of *Streptomyces fradiae* NRRL 2702. After selection with erythromycin, a mutant of *S. fradiae* was obtained with apparent constitutive adenine dimethylation. Fermentation beers from the mutant strains of *S. lincolnensis* and *S.*

fradiae were found to contain significantly higher concentrations of lincomycin and tylosin than the respective wild-type strains (Weisblum, 1982).

3. STRATEGIES FOR THE GENETIC MANIPULATION OF ANTIBIOTIC-PRODUCING MICROORGANISMS

In addition to the available natural genetic variation in wild strains of antibiotic-producing microorganisms, genomes of divergent biosynthetic capabilities may be constructed by laboratory gene manipulation. Techniques exist and continue to develop for modifying known antibiotics by feeding precursor analogues into blocked pathways, by mating strains capable of making hybrid products, or, increasingly, by cloning specific genes able to introduce specific enzymic or regulatory modifications to microbial products (Table 7.2). There is no reason that these techniques need to be restricted to known groups of metabolites. In situations in which there is a high probability of introducing genetic modifications, it should be feasible to screen the progeny for desired biological activity. The applications of genetics to industrial strains have been reviewed (Elander and Chang, 1979; Elander and Demain, 1981; Hopwood, 1981; Queener and Baltz, 1979).

The techniques for genetic modification include mutation, recombination, transformation–transduction, and gene cloning; often a combination of these is used. Mutants can be directed toward a number of objectives. They provide a basis for the selection of recombinants and are powerful tools in the elucidation of biosynthetic pathways. It is possible to delete nonessential structural or regulatory genes, perhaps concerned with the synthesis of one

TABLE 7.2 New Genetic Approaches for the Genetic Improvement of Microorganisms

Strategy

Generate biosynthetic mutants (idiotrophs) for precursor feeding (mutasynthesis)
Develop hybrid strains to produce hybrid antibiotics
Use gene cloning to redirect antibiotic biosynthesis

Techniques

Mutation and directed selection
Recombination by
   Conjugation
   Protoplast fusion
   Fungal sexual and parasexual cycles
Transformation, transfection, and transduction (enhance by using protoplasts)
Gene cloning by
   Target-directed cloning of specific genes
   Random shotgum cloning to generate gene libraries

bioactive metabolite, with effects on the type or quantity of metabolites produced. Mutants blocked in biosynthesis of a metabolite may generate modified metabolites when fed with precursor analogues.

Natural systems of gene exchange are available in most groups of antibiotic-producing microorganisms but are well defined in only a few species. Thus, the exploitation of conjugation, transformation, and transduction in prokaryotes and sexual or parasexual cycles in eukaryotes is limited in value to highly developed industrial strains and is not easily applied to new metabolites in strains of new interest.

Over the past decade, protoplast techniques initiated by Okanishi et al (1974), have been developed to improve the frequency of recombination within and between species. Protoplast formation, fusion, and regeneration have accelerated the development of gene manipulation in *Streptomyces* sp. and is gradually finding applications in streptosporangia and other rare actinomycetes. These techniques have also been developed in fungi. Thus, in *C. acremonium* intraspecific fusions gave rise to haploid recombinants with increased antibiotic yields (Hamlyn and Ball, 1979). Interspecific fusions have been reported between *Aspergillus* sp. and between *Penicillium* sp. (Peberdy, 1979).

Techniques for transformation–transfection of protoplasts by plasmids and phage have exploited the new techniques for gene cloning in actinomycetes. Shotgun techniques have been used to clone antibiotic-resistance genes in streptomycetes. These experiments demonstrate the potential to mobilize and obtain expression of selected genes within and between species. It is only a matter of time before the vector systems are applied to the structural and regulatory genes involved in antibiotic overproduction.

The shotgun techniques used to clone selectable markers into vector systems can be used to prepare gene libraries from which selected genes can be transferred to hosts as the techniques are developed for detecting the appropriate clone. The preparation of gene libraries from strains able to overproduce prolific numbers and types of antibiotics and other metabolites may provide a powerful tool for subsequent selective gene cloning of an interspecific of intergeneric nature.

3.1. Parasexual Recombination and Antibiotic Production in *Penicillium chrysogenum*

Some years ago, Sermonti (1959) carried out heterokaryon experiments between a low-producing auxotrophic strain, NRRL 1951 *pro,* and a high-producing strain, Wis. 49-133 *nic*. Homokaryotic segregants from the heterokaryon showed a clear association between the nuclear marker and yield, that is, *pro* segregants were low producers and *nic* segregants were high yielding. Thus, penicillin yield is determined by nuclear genes, and there is evidence from both heterokaryons and parasexual recombinants that several different genes are involved. Both the *y met* and *w ade* heterokryon of Elander (1967) and the "new hybrid" strain of Alikhanian and Borisova (1956) show significantly higher yields than their parent homokaryons, suggesting complementation between nonallelic genes.

Two phenomena suggest that genes determining increased yield or penicillin are recessive and that independently induced mutations could be allelic. When a diploid strain is derived via the parasexual cycle between a low-yielding and a high-yielding strain, its yield is comparable to that of the former. Diploids obtained between strains from a common ancestor, but carrying independently induced mutations for higher yield, gave yields equivalent to the parental strain but higher than the yield of the common ancestor (McGuire et al., 1980; Elander, 1967). The strains may have carried a number of allelic mutant sites in common, even though the sites had been independently mutated. It appears that there are genetic determinants, recessive in their expression and located on more than one chromosome, that determine increased penicillin yield in *P. chrysogenum*. Their expression can be modified by genetic background, since most conidial color and auxotrophic markers reduced yield drastically. To date, mutation and selection have been the most reliable procedures for improving penicillin titers. Breeding involves the parasexual cycle, and, therefore, the peculiarities of the cycle in *P. chrysogenum,* present difficulties in achieving recombination. Strain variability and instability cause difficulties in the initiation of the parasexual cycle because heterokaryons are not easily produced with strains of *P. chrysogenum*. When heterokaryons are produced, instability of diploids and their segregants often occurs. Elander (1967) described a highly stable diploid strain at Eli Lilly and Company derived from the haploid production strain E-15 (Figure 7.4). Spontaneous variation in the diploid and haploid strain was 9.2 and 31.9%, respectively, as assessed by the color types that segregated. The difference was even more striking following exposure to UV irradiation, in which case the diploid:haploid-derived variants were 11.6 and 41.1%, respectively. In this case, the diploid strain was more stable than its haploid progenitor. In contrast, Ball (1973) reported that diploids show instability greater than that of their haploid progenitors but within the same range. The diploid showed twice as many poorly sporulating types as did the haploid, whereas densely sporulating types were increased 10-fold. This phenomenon has been compared with that of "mitotic nonconformity" in *Aspergillus nidulans* (Nga and Roper, 1969). Mitotic nonconformity stems from the existence of duplicate segments of small fractions of the genome that probably originated through translocations. If the difficulties in producing diploids can be overcome, they may be attractive to the fungal geneticist. There is the possibility of a heterotic effect on yield. Even if this does not occur, productive segregants from diploid strains may be selected. If the latter approach is used, starting initially with strains lacking chromosomal rearrangements or inducing variants in increased yield from an existing strain without inducing chromosomal aberrations is recommended. Since spontaneous dediploidization can always occur, systems have been proposed to reduce such effects.

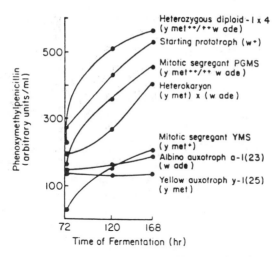

Figure 7.4. Penicillin production by mutant and recombinant strains of *P. chrysogenum* (From Elander, 1967.)

Macdonald and Holt (1976) have suggested the use of parental haploids that grow poorly in complex media to select against the parental segregants arising during fermentation. Azevedo and Roper (1967) suggested the induction of recessive lethals in the diploid so that haploid segregants would be eliminated. It has been suggested that the Lilly diploid may have been a balanced lethal diploid in that it was extremely stable but, even after treatment with UV irradiation, its proportion of viable segregants was only slightly increased. Provided diploid stability and unrestricted parasexual recombination can be achieved, it should be possible to progress further with selection. Elander (1967) described a spontaneous segregant from the Lilly diploid that produced nearly 25% more antibiotic than its parent, which itself yielded better than the production haploid E-15. Ball (1981) has also described recombinant segregants with improved yields.

The first 30 years of mutation selection and scientific breeding have demonstrated their potentials and the dangers of mutagens that induce chromosomal aberrations. The next 25 years should produce more extensive application of the parasexual cycle to breeding (Macdonald and Holt, 1976; Hopwood, 1977; Hopwood and Merrick, 1977; Elander, 1979).

3.2 Protoplast Fusion Technology

Gene transfer following the fusion of microbial protoplasts represents another approach to gene transfer. Using complementary biochemical mutants, protoplasts are generated by treating whole cells with a variety of lytic enzymes. An osmotic stabilizer is essential to provide osmotic support for the protoplasts, and fusion is enhanced by the addition of polyethyleneglycol (PEG) and magnesium ion. Reviews offer an excellent background on protoplast fusion methodology (Peberdy, 1979; Anne, 1976).

Hamlyn and Ball (1979) have used protoplast fusion technology with success with *C. acremonium,* an organism in which conventional genetic manipulations have proven to be difficult. They found recombinant strains that produced significantly more cephalosporin C than the biochemically deficient strains and that produced cephalosporins efficiently from inorganic sulfate.

Protoplast fusion technology has also been applied at Bristol-Myers Company for the development of fast-growing low *p*-hydroxypenicillin V strains of *P. chrysogenum* (Chang et al., 1982). Lowering *p*-hydroxypenicillin V was important because the hydroxylated product interferes with the chemical ring expansion of penicillins to oral cephalosporin products. Figure 7.5 shows the application of protoplast fusion technology to strains of *P. chrysogenum.*

3.3 Strain/Plasmid Interactions in Antibiotic Biosynthesis

The participation of plasmids in the biosynthesis of antibiotics was first pointed out by Okanishi et al. (1970). Plasmid genes probably participate in the formation of aerial hyphae, resistance to antibiotics, and other phenotypic characteristics. However, other interpretations of the data are also possible. Curing agents either induce or select spontaneous chromosomal rearrangements. Chromosomal reorganization could vary in certain parts of the chromosome and affect antibiotic production (Schrempf, 1982).

Streptomyces kasugaensis produces aureothricin, thiolutin, and kasugamycin. The production of all three antibiotics can be eliminated by the so-called curing treatment. Intensive study of aureothricin biosynthesis (Okanishi, 1979) showed that the structural genes for aureothricin biosynthesis seen to be localized on a chromosome, whereas plasmid genes may be included in the structure of the membrane and can affect the accumulation of precursors of the antibiotic in the cell. Another intensively studied model of interaction of plasmids with antibiotic synthesis is *Streptomyces venezuelae,* the producer of chloramphenicol (Akagawa et al., 1979). In this case all or a majority of structural genes responsible for antibiotic synthesis are located on the chromosome; the plasmid affects the increase in production by a mechanism that is still obscure.

The genetics of production of methylenomycin A and actinorhodin by *Streptomyces coelicolor* is understood in considerable detail. The genes for the synthesis of methylenomycin are born on the SCPl plasmid (Hopwood, 1979). On the other hand, actinorhodin is determined chromosomally (Hotta et al., 1977). The biosynthesis of aminoglycosidic antibiotics kanamycin, neomycin, and paromycin is assumed to involve the participation of a plasmid in the biosynthesis of deoxystreptamine, a common precursor of these

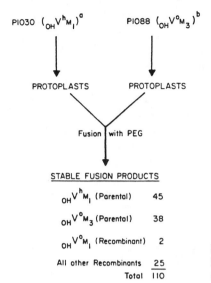

PIO30 $\left(_{OH}V^h M_1\right)^a$ PIO88 $\left(_{OH}V^o M_3\right)^b$

PROTOPLASTS   PROTOPLASTS

Fusion with PEG

STABLE FUSION PRODUCTS

$_{OH}V^h M_1$ (Parental)   45

$_{OH}V^o M_3$ (Parental)   38

$_{OH}V^o M_1$ (Recombinant)   2

All other Recombinonts   25

Total   110

Figure 7.5. Use of protoplast fusion for the breeding of low-p-hydroxypenicillin V strains of *P. chrysogenum*. (From Chang et al., 1982.)

antibiotics (Hotta et al., 1977). Results with istamycin indicate that plasmids could participate in the catabolite regulation by glucose and in membrane permeability. Plasmids also were studied in other producers of antibiotics of the neomycin series, in particular in connection with the so-called modifying enzymes–glycoside-3' -phosphotransferase and aminoglycoside-acetyltransferase, whose activities cause the resistance to or, more exactly inactivation of, these substances (Davies et al., 1980). The function of plasmids in producers of macrolide antibiotics has shown that plasmid genes play a role in the expression of antibiotic resistance and melanin formation (Schrempf, 1982). The high variability of the genes participating in melanin formation has been reported (Keiser et al., 1978).

An industrial strain of *Streptomyces kanamyceticus,* when propagated on a complex medium followed by plating, gave rise to a low frequency (0.2–1%) of small colonies having a soft fragmented texture that have lost their capacity to synthesize kanamycin (Km$^-$ phenotype) (Chang et al., 1980). Platings of sonicated mycelia grown in the presence of acridine orange (AO), ethidium bromide (EB), or acriflavin (Acr) resulted in a higher frequency (10–20%) of Km$^-$ colonies. Incubation of colonies at high temperature (35°C) accompanied by several serial transfers of Km$^+$ colonies also resulted in a high incidence (90%) of the Km$^-$ phenotype. The Km$^-$ colonies also developed from platings of conidia from streak cultures grown on agar containing AO, Acr, or EB where frequencies of 70–80% Km$^-$ colonies were routinely obtained. The Km$^+$ colonies were also found to be more resistant to kanamycin and amikacin (BB-K8) than Km$^-$ colonies, and the addition of kanamycin to oatmeal agar resulted in complete suppression of Km$^-$-type conidia. The Km$^+$ isolate also produced higher amylase activity in fermentation broths than the Km$^-$ isolates. Selected Km$^-$ isolates were found to produce kanamycin when grown in media supplemented with deoxystreptamine, streptamine, or D-glucosamine but not with streptidine; 2,6-dideoxystreptamine, 2,5,6-trideoxystreptamine; or 6-aminoglucose. Results were similar with other plasmid-cured strains of *S. kanamyceticus* (Hotta et al., 1977). These data are consistent with the hypothesis that active kanamycin production and other cultural characteristics may be controlled by plasmid gene(s) in *S. kanamyceticus* and that inclusion of kanamycin in growth media appears to maintain the integrity of cells containing plasmids, possibly by suppressing the growth of Km$^-$ cells. Incorporation of kanamycin in the vegetative inoculum stage also appears to stimulate kanamycin production (Figure 7.6).

3.4. Potentials for Recombinant-DNA Technology in Antibiotic Producing Microorganisms

In contrast to conventional mutation–selection methodology in which trial and error approaches are often used to generate superior microbial strains, recombinant DNA (R-DNA) technology results in a directed effort to create desired guided genetic changes (Elander, 1980). R-DNA technology singularly enables us to design genetically microbial strains with capabilities to synthesize end products that otherwise could not be generated. R-DNA technology has already been used to increase the synthesis of specific enzyme products over 500-fold (Hershfield, et al., 1974) and to generate novel hybrid antibiotics (Hopwood et al., 1985).

The basic requirements for the in vitro transfer and expression of foreign DNA in a host cell are outlined in Figure 7.7 and can be summarized as follows:

1. A vector DNA molecule (plasmid, phage, 2-μm DNA, etc.) capable of entering and replicating within a host cell. The vector should be small, easily prepared, and must contain one site in which the integration of foreign DNA will not destroy an essential function.
2. A method of splicing foreign DNA into the vector.
3. A method of introducing the hybrid DNA recombinants into the host cell and a discriminating procedure for selecting the presence of the foreign DNA.
4. A method for assaying for the foreign gene product.

R-DNA technology has been one of the most rapidly developing areas of science since its first demonstration in *Escherichia coli* by Cohen et al. (1973). Dramatic advances have appeared over the past 5 years from the first demonstration of the chemical synthesis, cloning, and functional expression

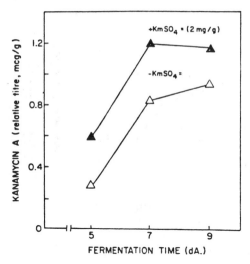

Figure 7.6. Effect of added kanamycin to inoculum stage on kanamycin production. (From Chang, unpublished data.)

of the human hormone somatostatin (Itakura et al., 1977) to the cloning and functional expression of synthetic human insulin in *E. coli* (Goeddel et al., 1979). Recombinant plasmids containing portions of human fibroblast interferon gene sequence have been reported in Japan (Taniguchi et al., 1979), and the synthesis of α, β, and γ human interferon related peptides has been described in bacteria and yeast (Derynck et al., 1982).

The successful demonstration of yeast transformation has resulted in a technique applicable to industrially important fungi for the isolation of

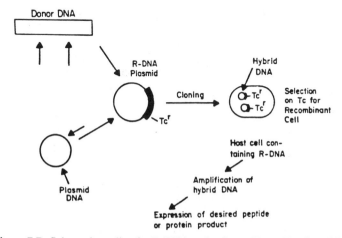

Figure 7.7. Schematic outline for R-DNA technology. (From Elander, 1981.)

specific yeast genes (Hinnen, 1978). Beggs (1978) has shown that the transformation involving a 2-μm yeast plasmid was successful in isolating de novo the LEU-2 gene of yeast. The yeast gene coding for arginine permease, CAN*I*, has been isolated from a pool of yeast DNA fragments inserted into YEp13 (Broach et al., 1979). Rapidly developing R-DNA technology now provides approaches that may result in the successful transfer of acyltransferase genes from *Penicillium* to *Cephalosporium*, thereby resulting in a strain synthesizing solvent-extractable cephalosporins (Queener and Baltz, 1979; Elander, 1980). The transfer of HABA acylase genes from butirosin-producing strains of *Bacillus circulans* to kanamycin-producing strains of *S. kanamyceticus* could lead to the direct synthesis of the important semisynthetic aminoglycoside antibiotic, amikacin.

R-DNA technology is also undergoing rapid development in streptomycetes, which produce 60% of the known antibiotics (Coats, 1982). Therefore, the development of DNA cloning and self-cloning systems (Figure 7.8) for *Streptomyces* would greatly facilitate the detailed genetic analysis for

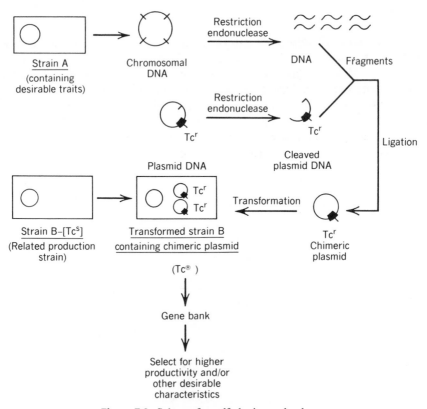

Figure 7.8. Scheme for self-cloning technology.

specific antibiotic biosynthetic pathways and of molecular mechanisms involved in their differentiation. Also, DNA cloning has resulted in new combinations of DNA gene sequences that produce novel antibiotics (Hopwood et al., 1985) and increased yields of existing or hybrid antibiotics. A potentially powerful DNA cloning system for interspecies gene transfer in a number of differing antibiotic-producing species of *Streptomyces* has been reported (Bibb et al., 1980). Thompson et al. (1982) have recently described newly constructed cloning vectors from streptomycete plasmid and genomic DNA termed pIJ 101 and pIJ 61, which provide for both replication and transfer functions. pIJ 101 is a multicopy (40-300/cell) plasmid with a broad host range occurring in *Streptomyces lividans* 5434. Moreover, the plasmid is self-transmissible and can mobilize chromosomal genes (Keiser et al., 1982). Katz et al. (1982, personal communication) has successfully cloned and expressed in protoplasts of *S. lividans* 1326 a DNA fragment using hybrid plasmids pIJ 700/701 coding for the synthesis of the enzyme tyrosinase from restriction digests of *Streptomyces antibioticus* DNA.

The applications of protoplast fusion, high efficiency DNA transformation, and R-DNA technology to industrially important bacteria, actinomycetes, and fungi appear numerous and will receive considerable attention over the next decade. Moreover, the new genetic technologies offer exciting prospects for the systematic genetic construction of improved microbial strains for the overproduction of existing and genetically modified "hybrid" antibiotics.

4. CONCLUSION

The application of rational (directed) selection techniques and genetic engineering methods for manipulation of antibiotic-producing microorganisms is generating a new era in industrial microbiology. Modern methods for increasing antibiotic productivity based on advances in the knowledge of the biosynthetic pathways and regulatory mechanisms for the induction and repression of genes involved in antibiotic synthesis are being described. The application of recombinant DNA and protoplast fusion methods to alter the genetics of antibiotic producers in a semirandom fashion for the development of novel hybrid antibiotics is being evaluated. Strategies for the specific alteration of antibiotic synthesis pathways by recombinant DNA techniques, where specific genes are introduced in order to add a small number of steps to existing pathways, are being discussed. Thus, rational mutation selection, protoplast fusion, and both semirandom and specific recombinant DNA methods have been illustrated in this chapter as alternative procedures for manipulating the biosynthetic pathways of microorganisms for strain improvement and for new hybrid antibiotic synthesis.

REFERENCES

Adelberg, E. Selection of bacterial mutants which excrete antagonists of antimetabolites. *J. Bacteriol.* **76,** 326, 1958.

Akagawa, H., Okanishi, M., and Umezawa, H. Genetics and biochemical studies of chloramphenicol-nonproducing mutants of *Streptomyces venezuelae* carrying plasmids. *J. Antibiol.* 32, 610–621, 1979.

Alikhanian, S. I., and Borisova, L. N. Vegetative hybridization of fungi of the genus *Penicillium. Izv. Akad. Nauk SSR.* 2, 74–85, 1956.

Anne, J. Somatic hybridization between *Penicillium* species after induced fusion of their protoplasts. *Agricultura* **25,** 1–117, 1976.

Arima, K. Recent developments and future direction of fermentation. *Jpn. Develop. Indust. Microbiol.* **18,** 79–117, 1977.

Azevedo, J. L., and Roper, J. A. Lethal mutations and balanced lethal systems in *Aspergillus nidulans. J. Gen. Microbiol.* **49,** 149–155, 1967.

Ball, C. The genetics of *Penicillium chrysogenum. Progr. Indust. Microbiol.* **12,** 47–72, 1973.

Ball, C. Genetic modification of filamentous fungi, in Smith, J. E., Berry, D. R., and Kristiansen B. (eds.), *Fungal Biotechnology.* London, Academic Press, 1981, pp. 43–54.

Ball, C., and Azevedo, J. L. Genetic instability in parasexual fungi, in Macdonald, K. D. (ed.). Second International Symposium on the Genetics of Industrial Microorganisms. London, Academic Press, pp. 243–251, 1976.

Beggs, J. D. Transformation of yeast by a replicating hybrid plasmid. *Nature (Lond.)* **275,** 104–109, 1978.

Bibb, M., Schottel, J. L., and Cohen, S. N. A DNA cloning system for interspecies gene transfer in antibiotic-producing *Streptomyces. Nature (Lond.)* **284,** 526–531, 1980.

Broach, J. R., Strathern, J. N., and Hicks, J. B. Transformation in yeast: Development of a hybrid cloning vector and isolation of the *CAN 1* gene. *Gene* 8, 121–133, 1979.

Chang, L. T., and Elander, R. P. Rational selection for improved cephalosporin C productivity in strains of *Acremonium chrysogenum. Develop. Indust. Microbiol.* **20,** 367–379, 1979.

Chang, L. T., Behr, D. A., and Elander, R. P. The effects of intercalating agents on kanamycin production and other phenotypic characteristics in *Streptomyces kanamyceticus. Develop. Indust. Microbiol.* **21,** 233–243, 1980.

Chang, L. T., Terasaka, D. T., and Elander, R. P. Protoplast fusion in industrial fungi. *Develop. Indust. Microbiol.* **23,** 21–29, 1982.

Coats, J. H. Models for genetic manipulation of actinomycetes. *Basic Life Sciences* **19,** 133–142, 1982.

Cohen, S. N., Chang, A. C. Y., Boyer, H. W., and Helling, R. B. Construction of biologically functional bacterial plasmids in vitro. *Proc. Natl. Acad. Sci., USA* **70,** 3240–3244, 1973.

Davies, J., Komatsu, K. I., and Leboul, J. Plasmids and modifying enzymes from aminoglycoside-producing organisms. Abstr. 6th Internat. Ferm. Symp. London, Ontario, American Soc. Microbiol., Washington, D.C., 1980.

Demain, A. L. New applications of microbial products. *Science* **219,** 709–714, 1983.

Derynck, R., Leung, D. W., Gray, P. W., and Goeddel, D. V. Human interferon γ is encoded by a single class of mRNA. *Nucl. Acids Res.* **10,** 3605–3615, 1982.

Elander, R. P. Enhanced penicillin synthesis in mutant and recombinant strains of *Penicillium chrysogenum,* in Stubbe, H. (ed.), *Induced Mutations and Their Utilization,* Berlin, Akademie-Verlag, 1967, pp. 403–423.

Elander, R. P. Mutations affecting antibiotic synthesis in fungi producing β-lactam antibiotics, in Sebek, O. K., and Laskin, A. I. (eds.), *Genetics of Industrial Microorganisms.* Amer. Soc. Microbiol., Washington, D. C., 1979, pp. 21–35.

Elander, R. P. New genetic approaches to industrially important fungi. *Biotechnol. Bioengineer.* **22** (suppl.) 1, 49–61, 1980.

Elander, R. P. Strain improvement programs in antibiotic-producing microorganisms: Present and future strategies. Moo-Young, M., Robinson, C. W., and Vezina, C. (eds.), *Advances in Biotechnology, Scientific and Engineering Principles,* New York, Pergamon Press, 1981, vol. I, pp. 3–9.

Elander, R. P., and Aoki, H. β-Lactam-producing microorganisms: Their biology and fermentation behavior, in Morin, R. B., and Gorman, M. (eds.), *The Chemistry and Biology of β-Lactam Antibiotics.* New York, Academic Press, 1982, vol. 3, pp. 83–153.

Elander, R. P., and Chang, L. T. Microbial culture selection. Peppler, H., and Perlman, D. (eds.), *Microbial Technology.* New York, Academic Press, 1979, vol. II, pp. 243–302.

Elander, R. P., and Demain, A. L. Genetics of microorganisms in relation to industrial requirements, in Rehm, H. J., and Reed, G. (eds.), *Biotechnology, Microbial Fundamentals.* Weinheim, Verlag Chemie, 1981, vol. I, pp. 237–277.

Elander, R. P., Mabe, J. A., Hamill, R. L., and Gorman, M. Biosynthesis of pyrrolnitrins by analogue-resistant mutants of *Pseudomonas fluorescens. Folia Microbiol.* **16,** 157–165, 1971.

Elander, R. P., Espenshade, M. A., Pathak, S. G., and Pan, C. H. The use of parasexual genetics in an industrial strain improvement program with *Penicillium chrysogenum,* in Vanek, Z., Hostalek, Z., and Cudlin, J. (eds.), *Genetics of Industrial Microorganisms,* Amsterdam, Elsevier, 1973, vol. II, pp. 239–254.

Fujisawa, Y., and Weisblum, B. A family of r-determinants in *Streptomyces* spp. that specifies inducible resistance to macrolide, lincosamide and streptogramin type B antibiotics. *J. Bacteriol.* **146,** 621–631, 1981.

Godfrey, O. W. Isolation of regulatory mutants of the aspartic and pyruvic acid families and their effect on antibiotic production. *Antimicrob. Ag. Chemother.* **4,** 73–79, 1973.

Goeddel, D. U., Kleid, D. G., Bolivar, F., Heyneker, H. L., Yansura, D. G., Crea, R., Hirose, T., Krazemski, A., Itakura, K., and Riggs, D. Expression in *Escherichia coli* of chemically synthesized genes for human insulin. *Proc. Natl. Acad. Sci., USA* **76,** 106–110, 1979.

Hamlyn, P. F., and Ball, C. F. Recombination studies with *Cephalosporium acremonium,* in Sebek, O. K., and Laskin, A. I. (eds.), *Genetics of Industrial Microorganisms,* Washington, D. C., Amer. Soc. Microbiol., 1979, pp. 185–196.

Hershfield, V., Boyer, H. W., Yanofsky, C., Lovett, M. A., and Helinski, D. R. Plasmid ColEl as a molecular vehicle for cloning and amplification of DNA. *Proc. Natl. Acad. Sci. USA* **71,** 3455–3459, 1974.

Hinnen, A., Hicks, J. B., and Fink, G. R. Transformation of yeast. *Proc. Natl. Acad. Sci., USA* **75,** 1929–1933, 1978.

Hopwood, D. A. Genetic recombination and strain improvement. *Devel. Indust. Microbiol.* **18,** 9–21, 1977.

Hopwood, D. A. Genetics of antibiotic production by actinomycetes. *J. Natl. Prod.* **42,** 596–621, 1979.

Hopwood, D. Possible applications of genetic recombination in the discovery of new antibiotics in actinomycetes, in Ninet, L., Bost, P. E., Bouanchaud, D. H., and Florent, J. (eds.), *The Future of Antibiotherapy and Antibiotic Research,* London, Academic Press, 1981, pp. 408–415.

Hopwood, D. A., and Merrick, M. J. Genetics of antibiotic formation. *Bacteriol. Rev.* **41,** 595–635, 1977.

Hopwood, D. A., Malpartida, F., Keiser, K. M., Ikeda, H., Duncan, J., Fujii, I., Rudd, B. A. M., Floss, H. G., and Omura, S. Production of "hybrid" antibiotics by genetic engineering. *Nature* 314, 642–644, 1985.

Hotta, K., Okami, Y., and Umezawa, H. Elimination of the ability of a kanamycin-producing strain to biosynthesize deoxystreptomine by acriflavin. *J. Antibiot.* **30,** 1146–1149, 1977.

Itakura, K., Hirose, T., Crea, R., Riggs, A. D., Heyneker, H. L., Bolivar, F., and Boyer, H. W. Expression in *Escherichia coli* of a chemically synthesized gene for the hormone somatostatin. *Science* **198**, 1056–1063, 1977.

Keiser, T., Hutter, R., and Sutter, R. Genetic characterization of albino mutants of *Streptomyces glaucescens*. Abst. 3rd Internat. Symp. on the Genetics of Industrial Microorganisms. Madison, Wis., 1978, p. 35.

Keiser, T., Lydiate, D. J., Wright, H. M., Thompson, C. J., and Hopwood, D. A. Molecular genetics of *Streptomyces* plasmids. Abstr. Fourth International Symposium on the Genetics of Industrial Microorganisms (GIM-82, Kyoto), P-III-22, 1982, p. 123.

Kitano, K., Nozaki, Y., and Imada, A. Strain improvement in carbapenem antibiotic production by *Streptomyces griseus* subsp. *cryophilus*. Abstracts. Fourth International Symposium on the Genetics of Industrial Microorganisms (GIM-82, Kyoto) 1982, p. 66, O-VI-7.

Macdonald, K. D., and Holt, G. Genetics of biosynthesis and overproduction of penicillin. *Sci. Prog. (Oxford)* **63**, 547–573, 1976.

McGuire, J. C., Glotfelty, G., and White, R. J. Use of cerulenin in strain improvement of the daunorubicin fermentation. *FEMS Microbiol. Lett.* **9**, 141–143, 1980.

Nga, B. H., and Roper, J. A. A system generating spontaneous intrachromosomal changes at merosis in *Aspergillus nidulans*. *Genet. Res.* **14**, 63–70, 1969.

Okanishi, M. Plasmids and antibiotic synthesis in streptomycetes. Sebek, O. K., and Laskin, A. I. (eds.), Genetics of Industrial Microorganism. Washington, D. C., Amer. Soc. Microbiology, 1979, pp. 134–140.

Okanishi, M., Ohta, T., and Umezawa, H. Possible control of aerial mycelium and antibiotic production in *Streptomyces* by episomic factors. *J. Antibiot.* **23**, 45–47, 1970.

Okanishi, M., Suzuki, K., and Umezawa, H. Formation and reversion of streptomycete protoplasts: Cultural condition and morphological study. *J. Gen. Microbiol.* **80**, 389–400, 1974.

Peberdy, J. F. Fungal protoplasts: isolation, reversion and fusion. *Ann. Rev. Microbiol.* **33**, 21–39, 1979.

Queener, S. W., and Baltz, R. H. Genetics of industrial microorganisms, in Perlman, D. (ed.), Annual Reports on Fermentation Processes. New York, Academic Press, 1979, vol. 3, pp. 5–45.

Schrempf, H. Role of plasmids in producers of macrolides. *J. Chem. Tech. Biotechnol.* **32**, 292–295, 1982.

Sermonti, G. Genetics of penicillin production. *Ann. N.Y. Acad. Sci.* **81**, 950–972, 1959.

Singh, A., and Sherman, F. Association of methionine requirement with methyl-mercury resistant mutants of yeast. *Nature (Lond.)* **274**, 227–229, 1974.

Taniguchi, T., Sakai, M., Fujii-Kuriyama, Y., Murumatsu, M., Kobayashi, S., and Sudo, T. Construction and identification of a bacterial plasmid containing the human fibroblast interferon gene sequence. *Proc. Jpn. Acad. Sci. B,* 1979, pp. 464–469.

Thompson, C. J., Ward, J., Keiser, T., Katz, E., and Hopwood, D. A. Streptomycete plasmid cloning vectors. Abstr. Fourth International Symposium on the Genetics of Industrial Microorganisms (GIM-82, Kyoto), O-VIII-6; 1982, p. 76.

Weisblum, B. MLS resistance in *Streptomyces* spp: Relation to antibiotic production. Abstracts, Fourth International Symposium on the Genetics of Industrial Microorganisms (GIM-82, Kyoto), 1982, p. 3.

8

CHALLENGES
AND OPPORTUNITIES
IN PRODUCT RECOVERY

GEORGES BELFORT

Professor of Chemical Engineering
Rensselaer Polytechnic Institute
Troy, New York

1. INTRODUCTION

Isolation and recovery of fermentation products are as important as and sometimes more expensive than the fermentation itself. In fact, a recent correlation by Dwyer (1984), shown in Figure 8.1, suggests that the selling price of various biologicals is inversely proportional to their concentration in the starting material. During the early days, conventional chemical engineering was used by the fermentation industry with little thought about the new opportunities and challenges. Clearly, production had highest priority, and innovative processes for separation and product recovery could come later.

Enzyme purification is particularly important because of increased use of enzymatic processes in the food, textile, chemical, brewing, detergent, pharmaceutical, agricultural, and even petroleum industries (Darbyshire, 1981). Developments in the gene-splicing or genetic engineering field have opened new doors for specialized separation processes for proteins. Other new developments such as the production of monoclonal antibodies from hybridoma cells have spurred the need for imaginative separation processes.

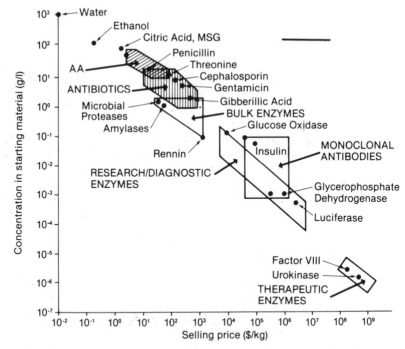

Figure 8.1. Concentration in starting material of various substances and their selling prices. (From Dwyer, J. L., 1984.)

Driven by economic considerations and the experience in the 1960s to produce single-cell proteins, continuous recycle fermentation systems are being developed to replace the usual batch fermentors. Higher concentrations of cells and/or products and the desire to recycle the cells without damaging them have introduced the need for new recovery processes. Other considerations include acceptable product purity, volume of production, and stability of the product to temperature, pressure, ionic strength, pH, and fluid shear.

Along with the recent spectacular advances in molecular biology has been the development of a wide range of new fractionation and concentration technologies. Besides improvements in the design and operation of traditional separations such as centrifugation, filtration, adsorption, ion exchange, and even sedimentation, recent advances in relatively new processes such as membrane separations, chromatography, and electrophoretic separations have expanded the options enormously.

This chapter will focus on enzyme recovery. Extraction and liberation of enzymes from whole cells by mechanical, physical, chemical, and biological disruption has been covered adequately in standard texts (Wang et al., 1979; Aiba et al., 1973) and will be omitted here. Darbyshire (1981) has indicated the essential differences between the isolation of extracellular and intracellular enzymes are (1) extracellular enzymes tend to be unstable and cannot easily be stored as crude supernatant, and (2) because they are present in relatively low concentrations, the extracellular enzymes should be concentrated before purification.

Finally, it should be emphasized that the fermentation and recovery steps are integral parts of an overall process such that yields and losses of products for each step must be considered in the context of the total system.

2. CONVENTIONAL FERMENTATION AND RECOVERY SYSTEM

Isolation of biochemicals begins with an extract of plant or animal tissues or with a fermentation broth. If the product has been released to the medium, cells, debris, and various solids are removed by filtration, sedimentation, or centrifugation. Products that are retained inside the cells must first be released, usually by rupturing the cells. This is followed by clarification usually using centrifugation. Recently the use of membranes in place of centrifugation for both cell separation and clarification has been analyzed (Hedman, 1984). A general scheme for downstream processing is shown in Figure 8.2.

Because the products are at low concentrations with large volumes of water present, the best option is some relatively inexpensive step with emphasis on concentration while achieving some purification. Unfortunately, some molecules are not easy to concentrate. Although evaporation will increase concentration of nonvolatile materials, energy is costly and there is little purification (separating volatile constituents has no value). Even when the product can be precipitated directly from the fermentation medium, costs tend to be unattractive and performance is not good. One practical example is precipitation of calcium salts of organic acids during the fermentation. This removes them and their inhibitory effects so that conversion of carbohydrate feedstocks is improved. The salts are treated with sulfuric acid to form calcium sulfate and free organic acids for further purification. Another example is the precipitation of the silver salt of the antibiotic cycloserine directly from the fermentation beer (broth after filtration). However, the silver salt is contaminated with other silver salts. The cost of silver is prohibitive and the silver cycloserine can explode, so better processes soon made this method obsolete.

Although there are superb methods for making quite pure enzymes, scale-up for large fermentation production is not yet practical. Crude proteases suitable for formulation in household detergents are prepared by adding acetone to the fermentation beer. Yields are poor because few enzymes have good stability in acetone. Nevertheless, other options for purification are not available.

When a molecule is soluble in organic solvents, extraction is ideal for rough purification. For example, penicillin has a carboxyl group that is rela-

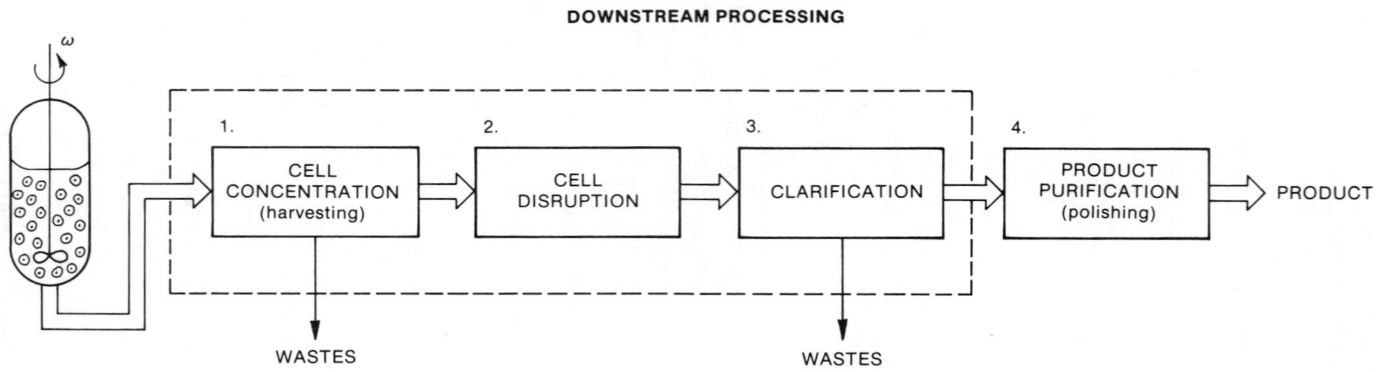

DOWNSTREAM PROCESSING

1. CELL CONCENTRATION (harvesting)

2. CELL DISRUPTION

3. CLARIFICATION

4. PRODUCT PURIFICATION (polishing)

WASTES

WASTES

PRODUCT

Figure 8.2. Downstream processing.

tively un-ionized at a pH below 5. Fermentation beer is acidified and very quickly contacted with organic solvent because stability is poor for penicillin at low pH. About five volumes of beer are contacted with one volume of solvent, and yields are good. Extraction back into an aqueous environment uses bicarbonate salts to avoid high pH, which is also disastrous to penicillin stability. Only molecules that are also solvent soluble at low pH and water soluble at higher pH follow the penicillin. The ratio of water to solvent during the back extraction is about 10:30, so the net concentration can be 100-fold. Purity of the penicillin in the aqueous extract is 70–80%, and this is sufficient for crystallization of the final product. Decolorization precedes final crystallization.

Many other molecules are soluble in solvents, but some very important biochemicals have a high affinity for water. Streptomycin is a sugary molecule that cannot be solvent extracted except by heroic means with special carriers. Enzymes and other proteins also defy solvent extraction. Fortunately, many biochemicals have ionizable groups that promote attraction to ion exchange resins. Elution with much smaller amounts of solutions than the large volumes of fermentation beers achieves good concentration, and there is excellent purification.

Recently, both solvent extraction and ion exchange have been tending to use whole broth without filtration. There are some problems due to the solids, but capital and operating costs are better and losses of filtration are avoided.

When neither solvent extraction nor ion exchange is feasible, the choices become difficult and expensive. One unattractive alternative is carbon adsorption. It is unselective and does not achieve much purification. Carbon adsorption was used initially for several different antibiotics but was supplanted by better methods. However, it may serve to get ready for other steps by reducing the volumes that must be handled and by providing a small degree of purification.

3. CONCENTRATION PROCESSES

An arbitrary division between those processes used to concentrate and those processes used to separate solids from liquids has been made. In Table 8.1 recovery and separation techniques used for extracellular and intracellular enzymes are summarized. In general the main objective is to dehydrate or remove water from the sample. In some cases contaminants are simultaneously removed resulting in partial purification, whereas in others concentration and purification are achieved in one step.

3.1. Precipitation

Selective removal or retention of a protein (enzyme) can be affected by adding a precipitant, thereby causing the protein to come out of solution.

TABLE 8.1. Recovery and Separation
Techniques

Precipitation
   Coagulation and flocculation
Differential migration
   Sedimentation
   Centrifugation
   Flotation
Filtration
   Deep Bed
   Cake
   Ultrafiltration
   Hyperfiltration
Sorption and partition
   Adsorption
   Ion exchange
   Affinity
   Gel filtration
   Liquid–liquid partition
Electrically driven
   Electrophoresis
   Electrofocusing
   Free-flow electropheresis
   Electrodialysis

Precipitation can be induced by the addition of neutral inorganic salts, organic solvents, charged flocculents, and even by changes in the pH. An example of the removal of protein with aluminun salts at different pHs is shown in Figure 8.3.

The salting out of proteins is a well-known concentration method. The observed decrease in protein solubility S is given by

$$\log S = a - bC \tag{8.1}$$

where C is the salt concentration and a and b are constants dependent on temperature and pH. Ammonium sulfate has been frequently used because it is inexpensive, highly soluble, and has a protective effect on proteins (Charm and Matteo, 1971). Addition of organic solvents at low temperatures ($-5°C$) reduces the dielectric constant of the water thereby reducing enzyme solubility. Denaturing of the enzyme has been noticed using this method, whereas toxic aerosols can be generated during recovery with centrifugation. Various polymer flocculents have also been used to induce precipitation of enzymes; however, because of high costs, this latter method has not found wide application in large-scale work. An exception has been polyethelene glycol (PEG), which is good for some high-value proteins.

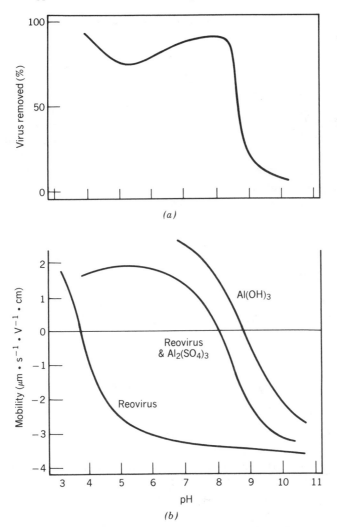

Figure 8.3. Comparison of virus removal and mobility for a coagulant at varying pH. (*a*) Loss of reovirus type 3 from suspension in 10^{-3} *M* NaCl and 8×10^{-5} *M* Al$_2$(SO$_4$)$_3$. C$_o$ (virus) $= 10^{10}$ ml^{-1}. (*b*) Mobility of Al(OH)$_3$, reovirus, and mixture as a function of pH. (After Taylor, 1980.)

To make the precipitation more manageable, additives such as starch, lactose, or kieselguhr have been used (Wang et al., 1979). Selective precipitants such as streptomycin sulphate, protamine sulphate, or manganese chloride have also been used to remove nucleic acids as contaminants in culture fluid.

Several techniques can be used to remove the enzyme precipitate. These include differential migration methods such as sedimentation, centrifugation, and flotation.

3.2. Coagulation and Flocculation

3.2.1. Stability

The process of aggregation of two colloidal species consists of two steps: (1) a transport step in which the species move toward each other and (2) an attachment step. The first step is termed flocculation, and the overall process is termed coagulation. Thermodynamically stable colloid systems do not spontaneously destabilize. Relatively stable colloid dispersions are unchanged over a period of days or weeks. Such systems are usually very sensitive to electrolytes that promote aggregation and sedimentation. The flocculating power of an electrolyte is dominated by the valence of the ion opposite in charge to that of the colloids.

3.2.2. Electrostatics

The double layer theory, based on the Debye-Hückel equation for strong electrolytes, gives

$$Kx = \ln \frac{[\exp(ze\phi/2kT) + 1][\exp(ze\phi_o/2kT) - 1]}{[\exp(ze\phi/2kT) - 1][\exp(ze\phi_o/2kT) + 1]} \qquad (8.2)$$

where

$$z = \text{valence}$$
$$e = \text{electronic charge}$$
$$\phi_o, \phi = \text{wall, field potential}$$
$$k = \text{Boltzmann's constant}$$
$$T = \text{absolute temperature}$$
$$x = \text{distance coordinate from the center of the colloid}$$

K^{-1} is called the "Debye length" and is given by

$$K = \left(\frac{2z^2e^2NC}{\epsilon kT}\right)^{1/2} \qquad (8.3)$$

where

$$N = \text{Avogadro's number} = 6.03 \times 10^{23} \text{ mol}^{-1}$$
$$e = \text{charge of electron} = 1.6 \times 10^{-19} \text{ C}$$
$$kT = 0.41 \times 10^{-20} \text{ V C at 293°K}$$
$$z = 1$$

Therefore,

$$K^{-1} = 2 \times 10^{-8} I^{-1/2} \text{cm} \qquad (8.4)$$

which suggests that an increase in the ionic strength, I, results in a decrease in K^{-1} and thus an increase in destabilization.

3.2.3. Electrostatic Colloid Stability
(Derjaguin-Landau-Verwey-Overbeek Theory)

Using the electrostatic models, the energy needed to bring a particle from infinity to some point x from a reference particle can be estimated. This is called the repulsive energy. By comparing this energy with the London or van der Waal's attractive dispersion energy between two equal particles, their stability can be calculated.

A simplified expression for the overlapping diffusion parts of the double layers gives (Reerink and Overbeek, 1954) the repulsive energy

$$V_R = \frac{B\epsilon k^2 T^2 a \gamma^2}{z^2} \exp(-KH) \tag{8.5}$$

where

$$
\begin{aligned}
B &= \text{constant} = 3.93 \times 10^{+39} \text{ A}^{-2} \text{ sec}^{-2} \\
\epsilon &= \text{permittivity} = 7.08 \times 10^{-12} \text{ C V}^{-1} \text{ cm}^{-1} \\
kT &= 0.41 \times 10^{-20} \text{ V C at } 20°C \\
\gamma &= \frac{\exp(ze\phi_\delta/2kT) - 1}{\exp(ze\phi_\delta/2kT) + 1} \\
z &= \text{valency of ions} \\
A &= \text{Hamaker's constant (depends on the atom density and} \\
&\quad \text{polarizability of the particles)} \\
\phi_\delta &= \text{Stern potential.}
\end{aligned}
\tag{8.6}
$$

For the case of two identical spheres of radius a where $a \gg H$ (Fig. 8.4), the London or van der Waals energy of attraction is given by

$$V_A = \frac{-Aa}{12H} \tag{8.7}$$

Clearly by adding the repulsive and attractive energy, a minimum of the net energy is at $dV/dH = 0$ and the net energy $V = 0$ when $V_R = V_A$. See Figure 8.5. Thus, from Eqs. (8.5) and (8.7)

$$V = V_R + V_A = \frac{B\epsilon k^2 T^2 a \gamma^2}{z^2} \exp(-KH) - \frac{Aa}{12H} = 0 \tag{8.8}$$

and

$$\frac{dV}{dH} = \frac{dV_R}{dH} + \frac{dV_A}{dH} = -KV_R - \frac{V_A}{H} = 0 \tag{8.9}$$

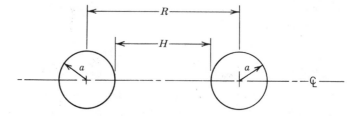

Figure 8.4. Particle coordinate system.

Solve for

$$H = K^{-1},$$ (8.10)

which on substitution into Eq. (8.8) gives

$$K = 4.415 \frac{B\epsilon k^2 T^2 \gamma^2}{A z^2}$$ (8.11)

Using this value of K (at $V = 0$) and the previous definition of K [after Eq. (8.3)], we get the molar concentration of an indifferent electrolyte:

$$c_o = \frac{9.75 B^2 \epsilon^3 k^5 T^5 \gamma^4}{A^2 e^2 N z^6}$$ (8.12)

We note that $c_o \propto z^{-6}$, that is, for a salt of valence 1, 2, 3, we need salt concentrations of $1^{-6}, 2^{-6}, 3^{-6}$, or 800:12:1, respectively, to effect the same removal. This is known as the "Schultz Hardy rule" and is shown in Figure 8.6.

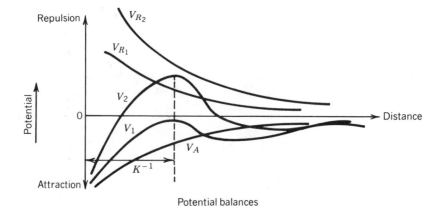

Potential balances

Figure 8.5. Potential profiles.

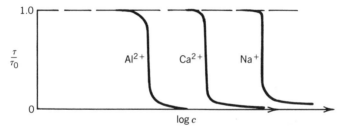

Figure 8.6. Schematic destabilization of silver iodide sols by addition of different electrolytes.

In addition to double-layer compression, colloids can be destabilized by adsorption and charge neutralization, enmeshment of a precipitate, and adsorption and interparticle bridging.

3.2.4. Kinetics

The kinetics of colloidal destabilization are classified into the following:

1. *Perikinetic flocculation,* which effects particle motion by Brownian diffusion as a result of "kT" energy. The concentration of colloids at time t is given by

$$N_t = \frac{N_0}{(1 + t/t_{1/2})} \tag{8.13}$$

where N_0 is the initial concentration of colloid (in units of number/ volume)

$$t_{1/2} = \frac{3\mu}{4\eta kTN_0} \tag{8.14}$$

where μ is the kinematic viscosity, η is the efficiency of attachment on collision, k is the Boltzmann's constant, and T is the absolute temperature.

2. *Orthokinetic flocculation,* which effects particle motion by physical aggitation of the fluid. The concentration of colloids at time t is given by

$$N_t = N_0 \exp\left(-\frac{4}{\pi}\eta\Omega\bar{G}t\right) \tag{8.15}$$

where $\Omega = [(\pi d_0^3/6)N_0]$ = volume fraction of colloids, d_0 is the original particle diameter of the monodispersed spheres, and $\bar{G}$ is the mean velocity gradient.

Comparing the rate of colloid removal between perikinetic and orthokinetic,

$$\frac{(dn_t/dt)_0}{(dn_t/dt)_p} = \frac{4\bar{G}\mu r_1^3}{kT} \tag{8.16}$$

which obtains for equal rates at 25°C in water:

$\bar{G}$ sec^{-1}	1	10	20	50	100
r_1 μm	1.05	0.5	0.4	0.3	0.23

Thus, the major result obtained from this analysis is that for particles smaller than about 0.1-μm, perikinetic flocculation dominates, whereas for particles larger than about 3-μm, orthokinetic flocculation dominates. The particle sizes between these ranges, that is, 0.1–3 μm, often pose real difficulties during filtration and sedimentation (Yao, 1968).

3.3. Differential Migration Processes

3.3.1. Sedimentation and Centrifugation

Settling phenomena have been divided into four distinct types. A description of each type is summarized in Table 8.2. Considering type 1 or discrete settling of microbial suspensions, the terminal velocity U_0 based on Stokes' Law of Resistance is given for gravitational field as

$$U_{o_g} = \frac{gd_p^2 (\rho_p - \rho_f)}{18\mu} \tag{8.17}$$

where ρ_p is the density of the microbial particle, ρ_f is the density of the fluid, d_p is the single particle diameter, and g is the acceleration due to gravity. For a centrifugal field

$$U_{o_c} = \frac{gZd_p^2 (\rho_p - \rho_f)}{18\mu} = AU_{og} \tag{8.18}$$

where Z is the effect of centrifugal force (i.e., $Z = r\omega^2/g$).

For type 2 or hindered settling of microbial suspensions, where the cell concentration has increased such that particle–particle interaction occurs, the velocity of settling is given by Aiba et al. (1973):

$$\frac{U}{U_{og}} = \frac{1}{1 + \alpha'\alpha^{1/3}} \tag{8.19}$$

TABLE 8.2. Type of Settling Phenomena

Type of Settling Phenomena	Description	Application/Occurrence
Discrete particle (Type 1)	Refers to the sedimentation of particles in a suspension of low solids concentration. Particles settle as individual entities, and there is no significant interaction with neighboring particles.	Very dilute cell suspensions or cells immobilized on spheres
Flocculent (Type 2)	Refers to a rather dilute suspension of particles that coalesce, or flocculate, during the sedimentation operation. By coalescing, the particles increase in mass and settle at a faster rate.	Cell concentration higher than in type I plus added flocculent
Hindered, also called zone (Type 3)	Refers to suspensions of intermediate concentration, in which interparticle forces are sufficient to hinder the settling of neighboring particles. The particles tend to remain in fixed positions with respect to each other, and the mass of particles settles as a unit. A solid–liquid interface develops at the top of the settling mass.	Typical for fermentation broth containing cells
Compression (Type 4)	Refers to settling in which the particles are of such concentration that a structure is formed and further settling can occur only by compression that takes place from the weight of the particles, which are constantly being added to the structure by sedimentation from the supernatant liquid.	Usually occurs in the lower layers of a deep cell mass, such as in the bottom of deep settling facilities and in thickening facilities

Source: After Metcalf and Eddy, 1979.

where

$$\alpha' = 1 + \bar{a}\,\alpha^{\bar{b}}$$

with $\bar{a} = 305$, $\bar{b} = 2.84$, $0.5 < \alpha < 0.15$ for irregular particles, and $\bar{a} = 229$, $\bar{b} = 3.43$, $0.5 < \alpha < 0,2$ for spherical particles. For dilute suspensions $\alpha' \sim 1$ to 2 for $\alpha < 0.15$. See Aiba et al. (1973) for further details on centrifuges. In some cases, addition of flocculent before sedimentation has resulted in vastly improved settling rates. See Figure 8.7 for several examples of this effect with changing pH and algal cells. Both the initial slope (kinetics) and the final turbidity will determine the choice of coagulant mixture most suitable for a particular solution.

LAMELLAR SETTLERS

Hazen (1904) realized that settler efficiency depended not on height but on surface area. He therefore suggested that "lamella" plates be inserted into a settler to obtain additional removal for type 1 settling. Boycott (1920) noted that red blood cells settled at a much higher rate in inclined tubes than in vertical ones. These two discoveries were then combined with recent mathematical analysis to give the current understanding of lamella settlers.

Acrivos and Herbolzheimer (1979, 1981) have analyzed inclined batch settlers and inclined continuous flow settlers. For infinite parallel flat plates, the settling has the form

$$\frac{dH}{dt} = -U_{o_g}\left(1 - \frac{H}{b}\sin\theta\right) + 0\,(\Lambda^{1/2}) \qquad (8.20)$$

where H is the height of the interface, b is the distance between plates, θ (90 $- \alpha$) is the angle of inclination from the horizontal, and Λ is the ratio of the Grashoff number to Reynolds number where $\Lambda \gg 1$. Their experiments showed that optimum removal rate occurred at $\theta = 60°$ with instability occurring at greater angles. See Figures 8.8 and 8.9.

Commercial inclined continuous lamellar settlers are already on the market and have been used successfully on activated sludge and trickling filter effluents (Slechta and Conley, 1971; van Vliet, 1971). Bungay and Millspaugh (1984) conducted laboratory experiments with a cross-flow lamellar settler for collecting yeast cells and primary sewage. They obtained 77 and 88% collection efficiencies for 1- and 2-hr detention times, respectively, for the yeast. For the sewage, 81 and 70–75% collection efficiencies for 1- and 6-hr detention times were measured.

3.4. Filtration

3.4.1. Deep Bed Filtration

CAPTURE AND ATTACHMENT IN ADSORPTION-ELUTION DEPTH FILTERS:
A THEORETICAL ANALYSIS

Mechanisms of Capture. The capture of fine particles such as viruses or cells in suspension by fitration through a porous medium may be divided into two steps: attachment and transport. Another step not well understood involves reentrainment or detachment.

Attachment. The surface characteristics of the virus or cell and the grain (or filter surface) determine directly whether attachment will occur. This is independent of the mechanisms of transport to the grain but can be influenced by the ionic strength of the aqueous solution. For the virus or cell to attach itself to the clean grain surface or previously adsorbed deposits, it has to be colloidally unstable. The classical theory of Derjaguin-Landau-Verwey-Overbeek (DLVO) describing colloidal destabilization can be used to study particle attachment (Derjaguin and Landau, 1941; Verwey and Overbeek, 1948). Electric repulsion forces are counterbalanced by London (van der Waals) dispersion attractive forces. As mentioned previously, mechanisms of attachment include charge neutralization, adsorption of charged polymers, and bridging (Weber, 1972). Fuhs et al. (1980) using a laser electrophoresis unit, were able to study the ζ-potential of reovirus and various floculents and clays as a function of pH (see Fig. 8.3). They explained virus attachment via the electrostatic model referred to above. Murray (1980) has also used the DLVO theory to explain virus association with soil particles.

Transport (after Ives, 1975). For aqueous filtration, the flow (rate) through porous media is usually laminar and proportional to pressure drop according to Darcy's law.

The dominant capture mechanisms are shown in Figure 8.10 and expressed quantitatively in Table 8.3. Although straining has been omitted, there is always the possibility that at the beginning of the run or as the filter nears exhaustion a cake may form at the top of the filter surface because of particle straining. This phenomenon of cake formation is most likely to occur on membrane-filter surfaces during the reverse osmosis and the ultrafiltration processes (Belfort and Marx, 1979). A dimensionless ratio is used to characterize each capture mechanism and to identify the important parameters.

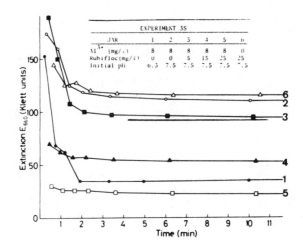

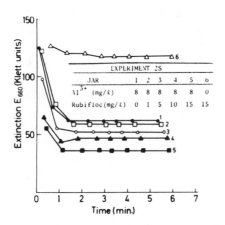

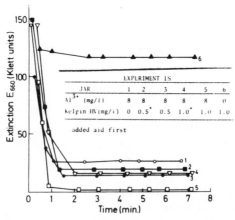

Figure 8.7. Settling curves for clean pond algal removal using alum and flocculant aids. (From Befort, unpublished, 1975.)

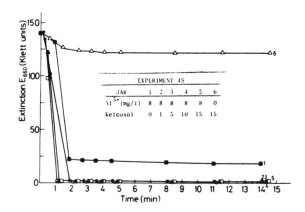

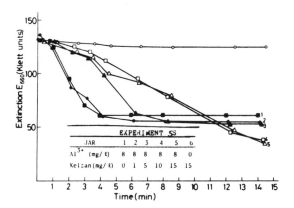

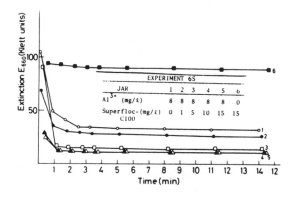

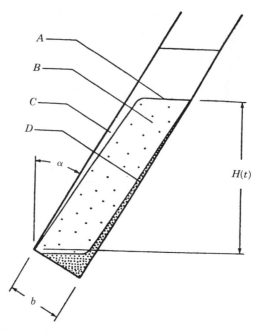

Figure 8.8. The different regions in the flow field: *A*, interface between the particle-free fluid and the suspension; *B*, suspension; *C*, particle-free fluid layer; *D*, concentrated sediment layer on upward-facing surface. (After Acrivos and Herbolzheimer, 1979).

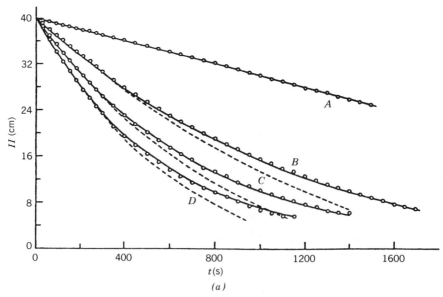

Figure 8.9. (*a*) Height of the top interface *H* versus time for $c_o = 0.10$; $H_o = 40$ cm and $b = 5$ cm ($\Lambda_o = 3.27 \times 10^7$, $R_o = 0.560$) for different angles of inclination α: *A*, $\alpha = 0°$; *B*, α 20°; *C*, $\alpha = 35°$; *D*, $\alpha = 50°$ ---, ordinary PNK theory; ———, PNK predictions accounting for the

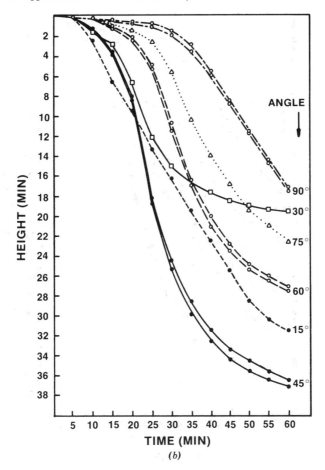

Figure 8.9 (*b*) (*continued*)

sediment layer. (After Acrivos and Herbolzheimer, 1979). (*b*) Whole blood sedimentation at different angles. Curves drawn from readings at 5-min intervals in Wintrobe tubes containing samples of the same blood. Note that the 45° angled tubes yield a greater drop in the 5 min of most rapid sedimenting, and that this period occurs early in the curve. (After Washburn and Meyers, 1957).

Efficiency of Particle Retention. The efficiency of particle retention has been expressed through the so-called filter coefficient defined by the Iwasaki rate equation:

$$\frac{\partial c}{\partial z} = -\lambda c \qquad (8.21)$$

where c is the particle bed concentration and z is the depth of the bed. The filter coefficient is related to the total collection efficiency η resulting from

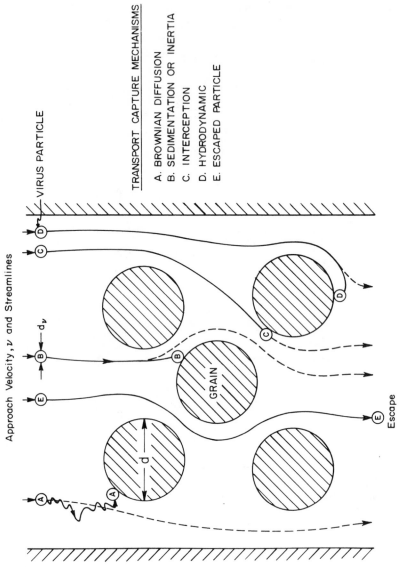

Figure 8.10. Transport capture mechanisms for particles passing through a depth filter.

TRANSPORT CAPTURE MECHANISMS

A. BROWNIAN DIFFUSION
B. SEDIMENTATION OR INERTIA
C. INTERCEPTION
D. HYDRODYNAMIC
E. ESCAPED PARTICLE

VIRUS PARTICLE

Approach Velocity, ν and Streamlines

d_ν

GRAIN

d

Escape

TABLE 8.3. Transport Capture Mechanisms in a Depth Filler

Mechanism	Relevant Equation	Relevant Dimensionless Number	Reference
A. Brownian diffusion	$D = kT/3\pi\mu d_v$	$N_{\text{diff}} = Pe^{-1} = \dfrac{D}{dV} = \dfrac{kT}{3\pi\mu dVd_v}$	Ives, 1975
B. Sedimentation	$V_s = \dfrac{g(\rho_v - \rho)d^2}{18\mu}$	$N_{\text{sed}} = \dfrac{V_s}{V} = \dfrac{g(\rho_v - \rho)d_v^2}{18\mu V}$	Ives, 1975
C. Inertia	—	$N_{\text{iner}} = \dfrac{\rho_s d_v^2 V}{18\mu d}$	Ives, 1975
D. Interception	—	$N_{\text{int}} = \dfrac{d_v}{d}$	Ives, 1975
E. Hydronamic	$V_l = \dfrac{K_1 V N_{\text{Re}}}{\epsilon}\left(\dfrac{d_v}{d}\right)^3 f(\eta')$	$N_{\text{hydro}} = \dfrac{V}{V_l} = \left[\dfrac{K_1 N_{\text{Re}}}{\epsilon}\left(\dfrac{d_v}{d}\right)^3 f(\eta')\right]^{-1}$	Altena and Belfort, 1984

Note: k = Boltzmann's constant; T = absolute temperature, °K; μ = viscosity, poise; d_v = particle diameter, cm; ρ_v = particle density, g/cm^3; ρ-fluid density, g/cm^3; V_l = lift velocity due to inertial migration, cm/sec; V = approach velocity, cm/sec; N_{Re} = axial Reynolds number based on grain diameter, dimensionless; d = grain diameter, cm; η = bed porosity, cm^3/cm^3; $f(\eta^1)$ = function whose value is between 0 and 1 and depends on η^1, the lateral position in a pore; D = particle diffusivity, cm^2/sec; V_s = sedimentation velocity, cm/sec.

all the mechanisms of capture (Rajaogopalan and Tien, 1979):

$$\lambda \simeq \left(\frac{3(1 - \epsilon)}{2d}\right)\eta \qquad (8.22)$$

where

$$\eta = \text{constant } N_{\text{diff}}^{\beta} N_{\text{sed}}^{\gamma} N_{\text{iner}}^{\xi} N_{\text{int}}^{\alpha} N_{\text{hydro}}^{\delta}$$

$$= \text{constant } \left(\frac{kT}{3\pi\mu dVd_v}\right)^{\beta}\left(\frac{g(\rho_v - \rho)d_v^2}{18\mu V}\right)^{\gamma}\left(\frac{\rho_s d_v^2}{18\mu d}\right)^{\xi}\left(\frac{d_v}{d}\right)^{\alpha}\left[\frac{\rho dV}{\epsilon\mu}\left(\frac{d_v}{d}\right)^3 f(\eta')\right]^{-\delta} \qquad (8.23)$$

where α, β, γ, δ, and ξ are positive exponents. Collecting items together,

$$\eta = \text{constant } \frac{d_v^a (V\rho_s)^{\xi}[\rho f(\eta')]^{-\delta}(kT)^{\beta}(\rho_v - \rho)^{\gamma}}{d^b\, {}^c V^d} \qquad (8.24)$$

where the exponents are given by

$$\begin{aligned}
a &= \alpha - \beta + 2(\gamma + \xi) - 3\delta \\
b &= \alpha + \beta - 2\delta + \xi \\
c &= \beta + \delta + \gamma - \xi \\
d &= \beta + \gamma + \delta
\end{aligned} \qquad (8.25)$$

Equations (8.24) and (8.25) indicate that η increases with the following: (1) Particle diameter d_v provided a is positive [i.e., $\alpha + 2(\gamma + \xi) > 3\delta + \beta$]. For very small particles, a may be negative and diffusion would dominate. (2) A decrease in grain diameter, d provided lift is negligible relative to the other forces (i.e., $\alpha + \beta + \xi > 2\delta$). (3) A decrease in viscosity, μ. (4) A decrease in approach velocity V.

In fact, experimental results show that for particles smaller than about 1 μm, Brownian diffusion (large β) is the dominant mechanism of removal, whereas for particles larger than about 5 μm, interception (large α) and sedimentation (large γ) are the dominant removal mechanisms (Yao, 1968). Inertial effects are generally very small in aqueous systems and can usually be ignored (small ξ). See Figure 8.11.

In designing an adsorption-elution system for the capture of very small particles such as viruses of diameter d_v, the maximum capture efficiency should occur at the highest fluid temperature tolerable by the virus [high $(kT)^\beta$ and low μ^c], at the lowest fluid approach velocity (low V^d), and with the smallest diameter capture surface or grain size (small d^b). Clearly, for completeness, a detailed attachment analysis incorporating the DLVO electrostatic theory is also necessary.

HIGH GRADIENT MAGNETIC FILTRATION

Nonmagnetic particles such as cells can be removed efficiently and quickly by coadsorbing these cells with a magnetic material such as magnetite in the presence of a flocculent such as ferric chloride. The mixture is then passed through a magnetic depth filter (Yadidia et al., 1977). The smaller the magnetite particles ($\sim 12~\mu$m) the greater the removal efficiency. A magnetic depth filter consists of a high-area magnetic surface, which when exposed to a magnetic field imposes high gradients of field. An example is a spaghetti sponge containing steel wire. Figures 8.12 and 8.13 illustrate the efficiency of removal of algal cells from laboratory-grown and pond effluent Scenedesmus algae. For economic viability, the magnetite and coagulation ($FeCl_3$) must be recycled and reused. The process removes over 90% of the algae (as measured by chlorophyll) for both feed solutions at much higher loading [about 24 gal/(ft^{2}min)] than sand filtration [about 2–6 gal/ft^2 min)].

3.4.2. Cake Filtration

Cake filtration involves the use of a septa (porous screen) onto which a precoat is deposited. The suspension is then filtered, with body feed added to reduce the compressibility of the cake. The cake itself represents the filtration media and captures additional particles as they are carried to the filter. Positive or negative (i.e., suction) pressure is used to drive the filtrate through the cake.

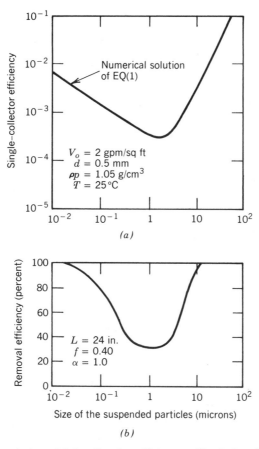

Figure 8.11. Theoretical model for filtration efficiency with single-collector and removal efficiencies as functions of the size of the suspended particles. (After Yao et al., 1970).

3.4.3. Microfiltration, Ultrafiltration, and Hyperfiltration

In situations in which depth and cake filters have been used to remove solid suspended particles from solution, microfiltration (MF), ultrafiltration (UF), and hyperfiltration (HF) are effective in removing particles, macro-molecules, and ionic molecules, respectively. Because of osmotic effects, HF (or reverse osmosis) requires a high driving pressure ($\sim$ 300–600 psi), whereas MF and UF only need relatively low driving pressure ($\sim$ 50–100 psi). Synthetic membranes are used to "filter" the particles, macro-molecules, or ions. The retained species concentrate on the up-stream side of the membrane because of the permselectivity of the membrane. The effect of this concentration polarization is to reduce the permeation flux through (1) increased osmotic back pressure, (2) gel or cake deposit offering a sec-ondary layer in series with the primary membrane, and (3) adsorption onto

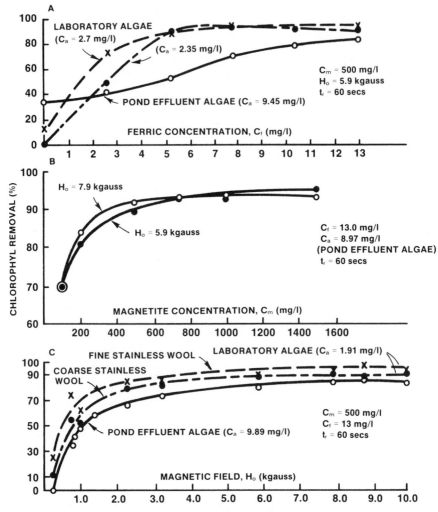

Figure 8.12. Chlorophyl/removal efficiency as a function of several process parameters for laboratory grown scenedesmus algae. (After Yadidia et al., 1977.)

the membrane surface. By propitious design, careful choice of membrane type, and use of optimum fluid management methods, concentration polarization effects can be reduced to economically acceptable levels. A detailed discussion on these approaches is dealt with in Chapter 10.

3.5. Chromatography

Chromatographic separations involve the retardation and separation of solutes from each other while being carried through a porous media by some

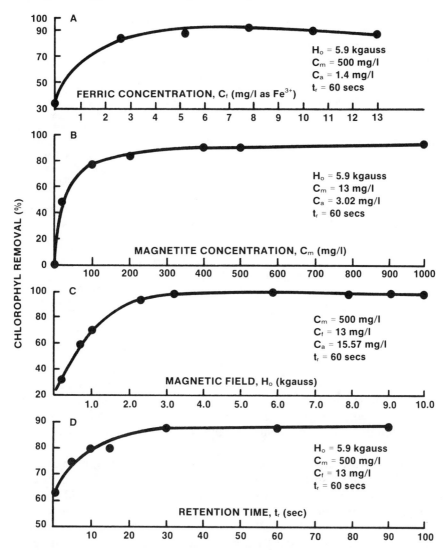

Figure 8.13. Chlorophyll removal efficiency as a function of several process parameters for laboratory and pond grown scenedesmus algae. (After Yadidia et al., 1977.)

carrier fluid. Several variant processes dependent on different dominant interactions are summarized in Table 8.4.

Adsorption chromatography has been used in the initial recovery of high-cost extracellular enzymes on a large scale. Batch techniques are generally used more than column techniques. In the latter case, pumping large volumes of supernatant through columns may result in inordinate pressure drops. Column techniques although slower than batch methods exhibit bet-

TABLE 8.4. Chromotographic Processes

Process	Dominant Interactions	Comments	Active Surface
Adsorption	Dispersion	Separates polar from nonpolar solutes	Alumina, silica gel, activated carbon
Ion exchange	Strong electrostatic (charged)	Separates polar solute as a function of ionic strength, pH, charge, and size	Amine, sulfonic, phosphoric acid groups attached to polystyrene or cellulose backbone
Affinity	Biospecific	Separates by biospecific interactions	Enzymes, antigens, antibodies, hormones attached or immobilized to silica or other carriers
Gel filtration	Hydrophilic plus low electrostatic	Uses hydrophilic gels with weak anion and cation exchanges to separate polar from nonpolar	Dextran, agarose, polyacrylamide
Liquid–liquid partition	Polar versus nonpolar	Partition coefficient	Polyethylene glycol and dextran or certain aqueous solutions

ter resolution. Langmurian isotherms are often used to describe the batch process.

The ion exchange process has been used for enzyme isolation; however, one disadvantage is the potential denaturation of the enzymes on exposure to the high charge density of the resins. However, ion exchange resins are easy to remove using sedimentation and because of their sphericity have good flow properties.

Since the hydrophilic gels (weak anion and cation) are less apt to denature enzymes and allow sorption and recovery under mild conditions, they are more attractive than strong ion exchange resins. Melling and Phillips (1975) give an excellent review of the basic principles of ion exchange procedures, the properties of the materials used, and their application to enzyme purification. Although over 90% of micrococcal nuclease is removed by SP-Sephadex C25 (weak hydrophilic) and phosphocellulose (strong ion exchanges), recovery by elution was 80 and 30%, respectively (Darbyshire, 1981).

A biospecific adsorbent is prepared by covalently coupling to a water-soluble substrate, a ligand specific to the enzyme, and a macromolecule of interest. The crude extract containing the desired enzyme is passed through a column allowing the enzyme and only the enzyme to be specifically adsorbed and removed from the solution. The required enzyme is then recovered by eluting with a solvent using ionic strength, temperature, pH, competitive inhibitor, or protein denaturants (urea, quanidine, or detergents). The major advantage of this approach is the excellent specificity for biologically active materials (Weetall, 1974). Ligand stability and fouling are some of the problems associated with this process.

Miller (1972) showed that the longer the chain that attaches the ligand to the substrate carrier, the better the enzyme is retained. Weetall (1974) gives an excellent review of some of the molecules that have been purified by affinity chromatography. His table is reproduced as Table 8.5.

Commercial carriers (solid supports) are available as celluloses, polystyrenes, polyamines, acrylics, cross-linked dextrans, agarose, and inorganic materials such as porous glass, all having functional groups to which various ligands can be attached. The major coupling methods include the use of carbodiimides, glutaraldehyde, hydrazides, cyanogen bromides, isothiocyanates, acid chlorides, alkyl halides, or diazonium salts.

Currently several bioengineering companies are using affinity methods in commercial production, although where possible ion exchange is used because of more favorable economics.

3.6. Electrically Driven Processes

Several processes in various stages of development and based primarily on the difference in electrophoretic mobility of a protein are currently being developed. These electrokinetic separation processes are summarized in Table 8.6. In addition to the processes listed in Table 8.6, newer processes such as forced flow electrophoresis, electrodecantation, and isotachophoresis are being scaled-up.

Besides electrodialysis, which has been used for large-scale desalting applications [$> 100,000$ gal/(ft^{2}day)], the other processes are not much larger than analytical and some preparative scale. Current efforts are being made to develop the other processes for large-scale application. Bier and Egen (1981) have developed a recycling isoelectric focusing apparatus that gives a rapid resolution of components differing by less than 0.2 pH units as shown by a number of fractionations of peptides, protein hormones, enzymes, snake venom factors, and blood protein.

3.7. Other Processes

Because of space, several important processes have not been represented. One that will only be mentioned because of its tremendous potential in the isolation of species of interest from complex mixtures of enzymes, proteins, nucleic acids, and cellular debris is two-phase aqueous partitioning. One phase contains a polymer such as PEG, whereas the other phase contains another polymer such as dextran. A very concentrated salt solution can substitute for one of the polymers, usually the dextran. The two polymers are chosen for their incompatibility. The resultant aqueous two-phase system will partition some species into one phase. Solids may aggregate at the interface. Proteins seldom have good partition coefficients, but sometimes

Macromolecule	Carrier	Liquid
Acetylcholinesterase	Agarose	[N-ε-Aminocaproyl)-p-aminophenyl] Trimethylammonium bromide
Acetylcholinesterase	Agarose	Substrate analogues
Anti-A phytohemagglutinin	Agarose	Blood group A substance
Avidin	Cellulose	Biotinyl chloride
Avidin	Agarose	Biocytin
Carboxypeptidase B (porcine)	Agarose	D-Ala-L-Arg
Carboxypeptidase A	Agarose	L-Tyrosine-D-tryptophan
Chymotrypsin	Agarose	ε-Aminocaproyl-D-tryptophan methyl ester
3-Deoxy-D-arabinohep-tulosonate-7-phosphate synthetase	Agarose	L-Tyrosine
Asparaginase	Agarose	Cross-linked dextran
Flavokinase	Cellulose	Flavins
β-Galactodidase	Cross-linked bovine γ-globulin	Paminophenylthiogalactoside

Macromolecule	Carrier	Ligand
Tyrosine aminotransferase	Agarose	Pyridoxamine
Trypsin	Aragose	Ovomucoid
β-Galactosidase	Agarose, polyacryl-amide	p-Aminophenylthiogalactoside
Glyceraldehyde-3-phosphate	Agarose	Adenosine monophosphate Nicotinamide adenine Dinucleotide
Glyceraldehyde-3-phosphate	Agarose	Adenosine monophosphate Nicotinamide adenine Dinucleotide
Glycerol-3-phosphate	Agarose	Halogenated guanisine Monophosphate
Hemoglobin	Agarose	p-Chloromercuribenzoate
Lactate dehydrogenase	Agarose	Oxamate
Lactate dehydrogenase	Agarose	Adenosine monophosphate Nicotinamide adenine Dinucleotide
Lactate dehydrogenase	Agarose	Adenosine monophosphate Nicotinamide adenine Dinucleotide
Alcohol dehydrogenase	Porous glass	Nicotinamide adenine dinucleotide
Mercaptopapain	Agarose	p-Aminophenylmercuric acetate
Amyliod protein (human)	Agarose	Congo red dye

TABLE 8.5 (*Continued*)

Macromolecule	Carrier	Ligand
Nuclease (staphyloccal)	Agarose, polyacryl- amide	*p*-Aminophenyl-pdt
Nuclease (staphyloccal) synthetic peptide of P_2	Agarose	P_3-Peptide
Papain	Agarose	Gly-Gly-(O-benzyl)-L-Tyr-L-Arg
Plasminogen	Agarose	L-Lysine
Proteins (sulfhydryl)	Cross-linked dextran	3,6-Bis(acetatomercurimethyl)- dioxane
Pyruvate kinase	Agarose	Adneosine monophosphate
Ribonuclease A (pancreatic)	Agarose	*p*-Aminophenyl p-U-cp
Ribonuclease-*S*-peptide (synthetic)	Agarose	S-Protein
Ribonuclease inhibitor (liver)	Carboxy- methyl cellulose	Ribonuclease
t-Ribonucleic acid (isoleuzyl)	Agarose	Specific tRNA synthetase
Thrombin	Agarose	*p*-Chlorobenzylamide-ε- aminocaproic acid
Tetrahydrofolate dehydrogenase	Agarose	Methotrexate
Tyrosinase	Cellulose	Aminophenol
Throxine binding protein	Agarose	L-Thyroxine
Trypsin	Agarose	Ovomucoid

Source: Wheetall, 1979.

[a] Immunoadsorbents for the isolation and purification of antigens and antibodies have not been included.

an excellent separation is achieved. The process is relatively efficient and very mild because surface tension between the phases is small. Scale-up to any size presents only minor engineering problems, but the polymers are expensive. Selectivity can be increased by coupling affinity ligands to the polymer. Current efforts are focused on developing a continuous extraction process (Brooks, 1983).

4. FUTURE

In this review, several well-known standard separation processes have been discussed. In addition, recent developments for newer processes with poten-

TABLE 8.6. Electrokinetic Separation Processes

Process	Electrical-Separation Based on	Bioapplications
Electrodialysis	Multicell pairs of anion and cation exchange membranes	Human plasma protein fractionation Fermentation mashes Desalting biofluids
Electrofiltration	Electrical potential balances and concentration polarization	Highly charged biomolecules
Electrophoresis	Combined gel-permeation and electrical potentials	Fractionation of proteins, DNA, etc.
Isoelectric focussing	pH gradient established by potential gradient	Protein purification Monoclonal antibody purification

tial in the biotechnology field have been reviewed. Exciting new processes such as magnetic filtration, affinity and gel filtration chromatography, electrofocusing, and electrophoresis all await scale-up for large-scale separations. Others such as the lamella settlers and membrane processes (ultrafiltration, hyperfiltration, and electrodialysis) have already been scaled-up but need to be evaluated for specific applications. Computer process control with continuous recycle systems will surely enter the biotechnology area soon.

REFERENCES

Acrivos, A., and Herbolzheimer, E. *J. Fluid Mech.* **92**, 435, 1979.

Aiba, S., Humphrey, A-E, and Mills, N. F. *Biochemical Engineering,* 2nd ed. Academic Press, New York, 1973.

Belfort, G., and Dziewulski, D. M. In Middlebrooks, E. J. (ed.). *Water Reuse,* Ann Arbor, Ann Arbor Science, 1982, Chap. 29, pp. 679–750.

Belfort, G., and Marx, B. *Desalination* **28**, 13–30, 1979.

Bier, M., and Egen, N. Large Scale Protein Purification by Isoelectric Focusing, presented at Advances in Fermentation Recovery Processes Techs. Banff, Canada, June 7–12, 1981.

Boycott, A. J. *Nature* **104**, 532, 1920.

Brooks, D. E. *Bio/Technology,* **1**, 668–669, 1983.

Bungay, H. R., and Millspaugh, M. P. *Biot. Bioengr.* **26**, 640–641, 1984.

Charm, S. E., and Matteo, C. C. *Methods Enzymol.* **22**, 476–556, 1971.

Darbyshire, J. Large-scale enzymes extraction and recovery. In Wiseman, A. (ed.), *Topics in Enzyme and Fermentation Biotechnology,* Chichester, Ellis Harwood, 1981, chap. 3, pp. 147–186.

Derjaguin, B. V., and Landau, L. D. *Acta Physiochim URSS,* **14**, 633, 1941.

Dwyer, J. L. *Bio/Technology* **2**(11) 957–964, 1984.

Fuhs, W., et al. Virus uptake by minerals and soils, presented at the 53rd Annual Conference of the Water Pollution Control Feder., Las Vegas, NV, Sept. 28–Oct. 3, 1980.

Green, G., and Belfort, G. *Desalination* **35**, 129–147, 1980.

Happle, J., and Brenner, H. *Low Reynolds Number Hydrodynamics*, Leyden, Noordhoff, 1973, p. 298.

Hazen, A. Transactions, *ASCE* **53**, 45, 1904.

Hedman, P. *Am. Biotech. Lab.* **2**(3), 29–39, 1984.

Helbagzheimer, E., and Acrivos, A. *J. Fluid Mech.* **198**, 485, 1981.

Ho, B. P., and Leal, L. G. *J. Fluid Mech.* **65**, 365, 1974.

Ives, K. J. Capture Mechanisms in Filtration, in Ives, K. J. (ed.), *The Scientific Basis of Filtration*, Alphen aanden Rijn, Noordloff, pp. 183–201, 1975.

Melling, J., and Phillips, B. W. In Wiseman, A. (ed.), *Handbook on Enzyme Biotechnology*, Chichester, Ellis Harwood, 1975, pp. 55–88, 181–202.

Metcalf and Eddy. *Waste-Water Engineering, Treatment, Disposal, Reuse.* New York, McGraw-Hill, 1979, p. 202.

Michaels, A. S. *Desalination* **35**, 329–351, 1980.

Miller, J. V. *Biochem. Biophys. Acta,* **276**, 407, 1972.

Murray, J. P. Physical chemistry of virus adsorption and degradation on inorganic surfaces, U.S. EPA Report–600/2-80-134, 1980.

Rajagoplan, R., and Tien, C. The theory of deep bed filtration in Wakeman, R. J. (ed.), *Progress in Filtration and Separation I,* Amsterdam, Elsevier, 1979.

Reerink, H., and Overbeek, J. Th. G. *Disc. Faraday Soc.* **18**, 74, 1954.

Slechta, A. F., and Conley, W. K. *JWPCF* **43**, 1724, 1971.

Taylor, D. H. Paper presented at 179th National ACS Mtg., March 24–29, Houston, TX, 1980.

van Vliet, B. M. *Water Res.* **11**, 783, 1977.

Verwey, E. J. W., and Overbeek, J. Th. G. *Theory of Stability of Lyophorsic Colloids*, Amsterdam, Elsevier, 1948.

Wang, D. I., Cooney, C. L., Demain, A. L., Dunnill, P., Humphrey, A. E., Lilly, M. D. *Fermentation and Enzyme Technology*, New York, Wiley, 1979.

Washburn, A. H., and Meyers, A. J. *J. Lab. Clin. Med.* **49**, 318–330, 1957.

Weber, W. J., Jr. *Physicochemical Processes for Water Quality Control*, New York, Wiley, pp. 67–75, 1972.

Weetall, H. H. (1974). Affinity chromatography, in Perry, E. S., van Oss, C. J., and Grushka, E. (eds.), *Separation and Purification Methods*. New York, Marcel Dekker, 1979, vol. 2, pp. 199–299.

Yadidia, R., Abeliovich, A., and Belfort, G. *Environ. Sci. Tech.* **11**(9), 913–916, 1977.

Yao, M. Influence of suspended particle size in the transport aspect of filtration, Ph.D. Dissertation, University of North Carolina, Chapel Hill, NC, 1968.

Yao, M., Habibian, M. T., and O'Melia, C. R. *Environ. Sci. Tech.* **5**(11), 1107, 1970.

9

SEPARATION
BY SORPTION

MICHAEL R. LADISCH

Professor of Agricultural and Chemical Engineering
Group Leader, Laboratory of Renewable Resource Engineering
Purdue University
West Lafayette, Indiana

Many processes for manufacturing biochemicals have costs that are dominated by the expense of purification. Fermentation products are diluted by water and contaminated by debris, salts, proteins, and a variety of compounds that may have properties quite similar to those of the desired material. Purification usually starts with some way to increase the concentration of the product so that large volumes of water need not be handled during the more selective steps. Processes such as solvent extraction and ion exchange can accomplish severalfold concentration and considerable purification. Separation of the product from molecules with similar properties can be very difficult. This chapter will cover two methods that are in large-scale use: column procedures for vapor-phase adsorption of water and liquid chromatography.

Liquid chromatography (LC) has gained a prominent position in separations of biological molecules during the last 20 years. This is particularly true for analytical-scale separations, where availability of the necessary instrumentation and chromatography supports has made this a widely used method (Regnier, 1983). Development of high-performance, preparative-scale LC separations is also proceeding. Systems capable of separating up to 1 kg/hr or more of product are available from several manufacturers.

The production of high fructose corn syrup can involve large-scale LC to upgrade fructose content from that attainable by enzyme conversion (typically 42–45%) by a partial separation of the glucose from the fructose (Antrim et al., 1979). A commercial system with countercurrent extraction scheme is the Universal Oil Products (UOP) Sarex Process (Wankat, 1982). The fructose is blended with high fructose corn syrup to obtain a 55% fructose product. This product is similar to cane sugar in sweetness and is used in many soft drinks.

Purification of antibiotics, amino acids, and other high-value products is amenable to LC separations. Engineering is needed to overcome the decreases in resolution, product concentration, and separation rate normally encountered in scaling up an LC separation. The factors that must be considered include the type of adsorbent or support, its packing characteristics, particle size, column length and diameter, and methods for ensuring uniform sample introduction and collection.

Another recent development is an adsorption process using corn to remove water from 190 proof ethanol vapors to obtain a substantially water-free product used as a gasoline octane booster. Distillation of ethanol from water still requires 15,000–30,000 Btu/gal. New drying technology to decrease this input is reducing ethanol cost.

The impact of energy cost for separating water from 2,3-butanediol is even greater. In this case, the discovery and development of an energy efficient separation technique for water from butanediol would greatly enhance the economic viability of obtaining this diol from renewable resources.

1. ADSORPTION USING POLYSACCHARIDES

The distillation of ethanol fermentation broth to 190 proof alcohol (92.4% by weight ethanol) is readily achieved (Katzen et al., 1980). Distillation above this concentration becomes more difficult and disproportionately more energy intensive because of the characteristics of the ethanol–water equilibrium curve and the existence of an azeotrope (95.6% by weight ethanol) at atmospheric pressure (Ladisch and Dyck, 1979). One approach to obtaining a substantially water-free product is to pass 190 proof vapor over an appropriate adsorbent (Garg and Auskiatus, 1983; Ladisch et al., 1984). A system using corn grits has recently been developed (Voloch et al., 1980; Hong et al., 1982; Ladisch et al., 1984).

The process consists of first drying corn grits packed in a stationary bed. Air, CO_2, or N_2 containing less than 0.015 mol fraction water is used at 80–100°C to dry the grits to a moisture of 2% or lower. Once the grits are dried and the adsorption bed is heated to 80°C or higher (i.e., above the dew point of 190 proof alcohol), adsorption is initiated. The 190 proof ethanol vapors are passed upflow through the column at a superficial velocity of 0.5–1 cm/sec. The water adsorbs on the corn grits, and the water-free ethanol vapors pass through the column. The heat of adsorption at these conditions is about 1200 Btu/lb water adsorbed (Rebar et al., 1984), and hence, a significant temperature rise occurs in the bed. At conditions of practical interest, the adsorption is characterized by a combined wave front where the temperature and concentration waves elute together (Ladisch et al., 1984a). Once breakthrough of the concentration (and temperature) waves begin to occur, the flow of vapor to the column is stopped.

Desorption is carried out by passing air, CO_2, or N_2 in a direction countercurrent to that used for adsorption. The moisture content of the regenerating gas must be less than 0.015 mol fraction water if satisfactory results are to be obtained. The gas is preheated to above 80°C before being passed through the column. Once the grits return to their initial moisture and temperature, the regeneration gas is shut off. The column is then ready for the next adsorption cycle.

Of particular interest is the selectivity of corn (separation factor of α = 1750) (Rebar et al., 1984), since only a negligible amount of ethanol is adsorbed (Hong et al., 1982). The polysaccharides in corn grits responsible for the adsorptive characteristics of the grits are starch, cellulose, and hemicellulose (Hong et al., 1982). Pure cellulose has been shown to have water sorption characteristics similar to starch. Dry biomass materials including agricultural residues and wood chips are also suitable adsorbents. These characteristics, together with the ability of corn to dehydrate other alcohols (Ladisch et al., 1984a; Ladisch and Tsao, 1982), suggest that polysaccharides may find unique applications in the separation of water from volatile fermentation products.

2. COLUMN DESIGN FACTORS IN LIQUID CHROMATOGRAPHY

The translation of a bench scale separation to a production scale must systematically consider column packing, sample introduction, dispersion phenomena, column capacity, particle size and velocity effects, and first estimation of separation costs. The discussion below of these key factors is based on experience with LC over spherical supports using water as the eluent. Parts are taken from a recently published review (Ladisch et al., 1984b).

2.1. Column Packing

Column packing materials and supports in large-scale use will typically be larger (40–300 μm diameter) than analytical scale supports (5–30 μm diameter). The size distribution of a commercial scale support (such as an ion

If $\rho > 1$.   **Figure 9.1.** Schematic illustration of settling that may occur during column packing.

exchange resin) is often broader than an analytical grade support. If the support has a density that is greater than that of the eluent (water), significant fractionation of the support can occur during the packing of a large column (e.g., 2 ft in diameter and 10 ft long). This situation (illustrated in an exaggerated manner in Fig. 9.1) is undesirable since it can cause sections of different void fractions to be formed. As a consequence, dispersion of a solute band moving down the column will be enhanced by the mixing caused by areas of different porosity in the bed.

Several measures can be taken to minimize fractionation of the packing material. The packing material can have a narrow size distribution. Attempts could also be made during support manufacture to control the density of the support to be close to that of the liquid used during the packing procedure. Minimizing differences in the rates of settling of different size particles will minimize fractionation. The effect of particle size on the terminal velocity, v_t, of a particle in a viscous fluid is illustrated by (Bird et al. 1960):

$$v_t = \frac{2R^2(\rho_s - \rho)g}{9\mu} \qquad \text{for} \qquad \frac{Dv_t\rho}{\mu} < 0.1 \qquad (9.1)$$

where R is particle radius; ρ_s and ρ are the densities of the particle and fluid, respectively, and μ is the viscosity. The settling velocity is proportional to the square of the radius. Hence, if the difference, $\rho_s - \rho$, is significant, significant differences in v_t will occur for particles having different values of R.

The density of the liquid could be adjusted (by temperature changes and dissolution of solutes) to be close to the density of the particles. Although these approaches may be feasible for aqueous systems where ion exchange resins ($\rho \cong 1$.) are used, inorganic-based supports with a relatively high density require very rapid pumping of support slurries into the column to minimize fractionation due to settling effects. Although this procedure is workable for analytical or preparative scale columns, it is more difficult to do on a commercial scale given the large volumes required.

2.2. Column: Particle Diameter Ratio

The average, particle diameter for supports used on an analytical scale typically range from 5 to 30 μm. The column diameter to particle diameter ratios for analytical columns inside diameter (i.d.) of 2–8 mm are on the order of 100–300. As the separation is scaled-up (in the case of ion exchange type supports), the average particle size also tends to be larger (40–300 μm) due to operating limits on pressure drop as well as support cost.

The cost of a commercial chromatographic grade resin is about $5–50/kg. In comparison, an analytical type resin costs about $100–4000/kg (anonymous, 1984). The analytical resin costs reflect the much smaller particle size (less than 20 μm) and narrower particle size distributions for these resins as compared to a commercial grade resin.

The choice of column diameter is important in view of known differences in void fraction in the packing of a spherical support material as a function of radial distance from the column wall (Cohen and Metzner, 1981). The void fraction at the wall is 1. and then decreases to the average value of the bed (typically in the range of 0.3–0.5) as the center of the column is approached. At least 40 particle diameters (away from the wall) are required for an average value to be attained. The higher void fraction near the wall allows the possibility of sample "fingering," with sample movement in the axial direction being more likely to occur along the walls of the column than in the bulk of the column packing. This, in effect, is another form of dispersion that decreases column efficiency since the sample no longer resembles a plug.

2.3. Sample Loading

After the column has been packed and equilibrated with the eluent, the sample is loaded. The loading of the sample must be carried out in a manner that allows the sample volume to be introduced as a plug with as little mixing with the eluent as possible. This is facilitated by a properly designed distributor consisting of a porous plate that has a sufficient pressure drop to distribute an incoming liquid feed evenly (see Fig. 9.2). The distributor plate is situated directly upon the column packing material to minimize mixing of the sample with the eluant, which would otherwise occur if a dead volume were present. Again, at the outlet the objective is to minimize mixing due to extra column effects. Extra column effects for analytical systems are discussed elsewhere (DiCesare et al., 1981). These effects also impact the efficiency of preparative and commercial scale separation systems.

SAMPLE DISTRIBUTION
(Minimize Dispersion!)

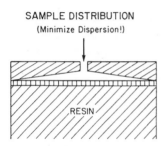

RESIN

Figure 9.2. Schematic representation of proper placement of feed distributor to minimize dispersion.

2.4. Dispersion: The Concept of Plate Height

Once the sample is introduced into the column (preferably as a "plug"), the tailing edge is washed through the distributor and onto the column by the eluant. A sample having a volume of 0.01–0.5% of the column void volume is typical of analytical scale chromatography for which the following analysis is applicable.

A sample plug pushed through the column by the eluant will have a tendency to disperse. As it elutes, it may give a gaussian-shaped peak as illustrated in Figure 9.3. The *average* solute concentration (i.e., total solute–eluant volume corresponding to the peak width) is typically reduced by a factor of 3–20-fold relative to the inlet sample concentrations. The width of this peak at the base is 4σ, and at half-height is 2.35σ. The number of plates, defined in terms of σ is

$$N = \frac{L}{H} = \frac{L^2}{\sigma^2} \tag{9.2}$$

where $\sigma = \sqrt{HL} = W/4$, N is the plate count, and H is the plate height. L is usually defined as the column length, although this is not strictly correct

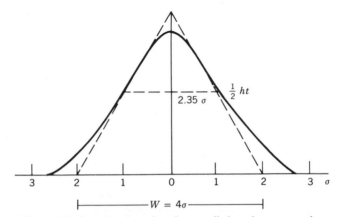

Figure 9.3. Gaussian-shaped peak as applied to chromatography.

(Martin and Synge, 1941; Knox, 1973; Jacobson, 1982). Rearrangement and expression in terms of t_r, the time required for the peak to elute after the sample is injected, and t_w, the peak width in units of time, gives

$$N = \frac{L^2}{(W/4)^2} = 16 \left(\frac{t_r}{t_w} \right)^2 \tag{9.3}$$

Since it is sometimes difficult to measure the peak width at the base, the width at half-height ($t_{w,1/2}$) is used instead (at half-height, $W = 2.35\sigma$). The plate height, H, is obtained from 9.4 by dividing the length of the column:

$$H = \frac{L}{N} = \frac{L t_{w,1/2}^2}{5.54 t_r^2} \tag{9.4}$$

A high plate count (or small plate height) is not necessarily synonymous with good resolution. Indeed, commercial scale columns with observed plate counts of 50–100 can give close to baseline separation of two- and three-component mixtures. A good separation depends on the combined factors of (1) plate height (a measure of sample dispersion inside the column) and (2) differences in the capacity of the packing to dynamically absorb or retain (and then desorb) two or more components as they pass through the column (i.e., a difference in capacity factors).

2.5. Capacity Factor (k')

The capacity factor, k_1', for a single component (i.e., solute 1) is defined as (Knox, 1973)

$$k_1' = \frac{\text{Moles of solute 1 in stationary phase}}{\text{Moles of solute 1 in mobile phase}} = K \frac{V_s}{V_m} = \frac{X_s V_s}{X_m V_m} \tag{9.5}$$

where V_s and V_m are the volumes of the sample in the stationary and mobile phases, respectively, X_s and X_m are the concentrations of the solute in the stationary and mobile phases, respectively, and K is the distribution coefficient between the two phases.

The elution of a solute is characterized by the retention volume, V_r,

$$V_r = V_m + K V_s \tag{9.6}$$

When Eq.(9.6) is combined with Eq. (9.5), it gives

$$k' = \frac{V_r - V_m}{V_m} \tag{9.7}$$

where V_r is the retention volume of the solute peak (corresponds to retention time, t_r) and V_m is equivalent to the column void volume (V_0) (measured by retention time, t_0, of an excluded component). Consequently, the relation for the capacity factor becomes

$$k' = \frac{t_r - t_0}{t_0} \tag{9.8}$$

The void volume, V_0, can be determined by injecting blue dextran (M.W. = 2,000,000), which is excluded from penetrating most ion exchange supports because of its size.

The capacity factor, k', is often taken as an intrinsic property of the support. In fact, it can be shown that k' is a function of ϵ, the column void fraction, as well as K, the distribution coefficient (which is a property of the support) (Jacobson, 1982; Ghin and Chang, 1982). It can also be shown that k' will increase with decreasing void fractions. This provides some rationale for the increase in resolution obtained in systems that mechanically compact the support after packing. An example is the Elf Aquitaine Series 300 LC (anonymous, 1983) in which a piston is used to compress the packing material in the axial direction. Good resolution of 0.80 kg of androstenedione from 0.20 kg of androstadienedone is reported over a column containing 40 kg of Merck silica H 60 at a linear eluent velocity of about 0.70 cm/min. The separation requires 6–7 hr.

2.6. Resolution

The resolution factor, R_s, is a measure of how well two bands are resolved. It is defined and calculated from the equation

$$R_s = \frac{(t_{r2} - t_{r1})}{\frac{1}{2}(t_{w1} + t_{w2})} \tag{9.9}$$

where t_{r2} and t_{r1} are the retention times of the bands (solutes in units of time, and t_{w1} and t_{w2} are the peak widths at baseline of these bands in units of time.

According to Snyder and Kirkland (1979), the capacity factor, the number of theoretical plates, and the resolution factor are related by

$$R_s = \frac{1}{4}(\alpha - 1)\sqrt{N}\left(\frac{k'}{1 + k'}\right) \tag{9.10}$$

This relationship is derived as follows (for bands 1 and 2) from Eq. (9.8):

$$t_{r1} = t_0(1 + k_1') \tag{9.11}$$

and

$$t_{r2} = t_0 (1 + k_2')$$ (9.12)

Assuming that $t_{w1} = t_{w2}$ and substituting these relationships into Eq. (9.9) gives

$$R_s = \frac{t_0 (k_2' - k_1')}{t_{w1}}$$ (9.13)

Since

$$N = 16 \left(\frac{t_r}{t_w}\right)^2$$ (9.14)

for band 1,

$$t_{w1} = \frac{4t_{r1}}{\sqrt{N}} = \frac{4t_0 (1 + k_1')}{\sqrt{N}}$$ (9.15)

Substituting this result into Eq. (9.13) gives

$$R_s = \frac{(k_2' - k_1') \sqrt{N}}{4(1 + k_1')} = \frac{1}{4} \left(\frac{k_2'}{k_1'} - 1\right) \sqrt{N} \left(\frac{k_1'}{1 + k_1'}\right)$$ (9.16)

The definition of the separation factor $\alpha = k_2'/k_1'$ thus results in Eq. (9.10), where k' represents the average values of the capacity factors of the two bands and where N in Eq. (9.16) represents the simple average of the plate counts for components 1 and 2. A resolution factor of 0.9 or higher gives a first indication of a suitable separation.

An example of a modest plate count, combined with a support having the appropriate capacity factors for two components, is given below. Let $N = 80$ and $k_1' = 1.25$ and $k_2' = 2.1$ (which gives $k' = 1.675$). In this case, a first approximation of resolution can be obtained from Eq. (9.10). Substituting in the values gives

$$R_s = \frac{[(2.1/1.25) - 1]}{4} \sqrt{80} \left(\frac{1.675}{1 + 1.675}\right) = 0.952$$

A resolution of 0.90 or higher typically indicates acceptable performance on a large scale. This numerical example illustrates that the absence of a high plate count (1000–50,000/m) normally associated with analytical columns does not necessarily indicate a poor resolution of the two components.

2.7. Scale-Up: Particle Size, Eluent Rate, and Sample Size

The diffusion coefficient, D, affects separation efficiency if this parameter is significantly different for the solutes being separated. Pieri et al suggest use of the Wilke-Chang equation to estimate D:

$$D = \frac{7.4 \times 10^{-12} \, T\sqrt{\psi M_{\text{eluant}}}}{\eta V_{\text{solute}}^{0.6}} \qquad (9.17)$$

where T is temperature (°K), ψ is the eluent association factor; M_{eluent} is the eluent molecular weight in grams, η is the eluent viscosity in MPa sec, and V_{solute} is the solute molar volume. In many LC separations the solutes being fractionated are similar. For example, in a phermone separation, two of the components are diastereomers, and the other solutes have similar values of D (Pieri et al., 1983). Based on their experience with the reverse phase separation of pheromones, Pieri et al. (1983) report that the key scale-up parameters are, in fact, the operational variables of particle size, eluent rate, and sample volume.

For linear chromatography (i.e., sample concentration is in a range that corresponds to a linear isotherm), the relationship between the maximum sample volume that can be injected, $V_{s,i}$, and the volume of the mobile phase inside the column, V_m (i.e., void volume) for $R_s > 1.3$ is (Gareil et al., 1983) as follows:

$$V_{s,i} = V_m \left(k_1' \, (\alpha - 1) - \frac{2}{\sqrt{N}} (2 + k_1' + \alpha k_1') \right) \qquad (9.18)$$

where $\alpha = k_2'/k_1'$ and the other parameters are as defined previously for an analytical system. Thus, determination of k_1', k_2', N, and V_m from an analytical injection allows estimation of $V_{s,i}$. For $R_s = 1.$, Pieri et al. (1983) present the expression

$$V_{s,i} = V_m \left((\alpha - 1)k_1' - \frac{1.25}{\sqrt{N}} (2 + k_1' + \alpha k_1') \right) \qquad (9.19)$$

Thus, for example, if $\alpha = 1.675$ for $k_1' = 1.25$, $k_2' = 2.1$, and $N = 80$, the maximum allowable sample volume, $V_{s,i}$, for $R_s \cong 1$ would be estimated at

$$V_{s,i} = V_m \left((1.675 - 1)(1.25) - \frac{1.25}{\sqrt{80}} (2 + 1.25 + 2.1) \right) = 0.0961 \, V_m$$

Equations (9.18) or (9.19) will predict that $V_{s,i} > V_m$ if k_1, k_2, and α are large enough. The maximum sample injection volume can be larger than the void volume if the adsorption of one component is so strong as to be almost irreversible. Elution of the adsorbed component may then require a different eluant or different operating conditions, which will cause the component to desorb. In the limit, the separation can then be characterized as an adsorption process with loading and regeneration cycles rather than as liquid chromatography.

If the second term on the right-hand side of Eqs. (9.18) and (9.19) is large, these equations reduce to

$$V_{s,i} = V_m k'_1 (\alpha - 1) \qquad (9.20)$$

This will typically occur if the $N > 1000$.

A material balance gives the amount of product Q_{s1} recovered. By definition, Eqs. (9.18)–(9.20) are developed for $R_s \geq 1$. Hence, all of the solute (for component n given by $Q_{s,n}$) originally injected is theoretically recovered, and $Q_{s,n}$ is simply

$$Q_{s,n} = C_{s,n} V_{s,i} \qquad (9.21)$$

where $C_{s,n}$ is the solute concentration of component n in the sample injected. This result is of limited use if full recovery of a solute is not achieved. In practice, recovery and purity will be determined experimentally, with relationships of the type given in Eqs. (9.19) and (9.20) giving only a first indication of the sample loading that might be possible. The treatment of deviations from ideal conditions, often encountered in production scale LC, is discussed later in this chapter.

The development of a process-scale LC separation often results from analytical scale data obtained under analytical scale conditions. Hence, the average particle diameter may be quite small (less than 20 μm) relative to those practical on a larger scale (> 50 μm). Factors that currently limit use of smaller particle sizes on a large scale include pressure drop limitations and untested operational stability under industrial conditions. However, these limitations are minor relative to the current lack of availability of commercial quantities of appropriately sized supports at a reasonable cost, a situation that is expected to change in the near future.

Current practice is to increase particle size as the scale of separation increases. This must be considered in estimating column productivity, since the plate count is a function of both particle size and the linear velocity of the eluant (Pieri et al., 1983). The relationship is given by

$$N = \frac{D^n}{m} \cdot \frac{L}{u^n d_p^{1+n}} \qquad (9.22)$$

where n, m are empirical constants, D the diffusion coefficient, L is the column length, u is the eluant linear velocity, and d_p is the particle diameter.

This result is obtained from the Snyder equation

$$h = mv^n \qquad (9.23)$$

where

$$h = \frac{L}{Nd_p} \qquad (9.24)$$

and

$$v^n = \left(\frac{ud_p}{D}\right)^n \qquad (9.25)$$

Combining Eqs. (9.23)–(9.25) and rearranging gives Eq. (9.22). Values of n in Eq. (9.23) range from 0.4 to 0.6, with 0.5 assumed by Pieri et al. (1983). If the plate height is specified to be constant and the particle size increases upon scale-up in a specified manner, either column length and/or eluant velocity must be adjusted. For example, let $N = 80$ with suitable analytical scale results being obtained for $d_{p,A} \cong 50$ μm at $u = 2$ cm/min and $L = 60$ cm. A large-scale column is to be packed with the same type of support, except that $d_p \cong 200$ μm $(= 0.02$ cm$)$. Assuming $n = 0.5$, where N for the analytical and large-scale columns are equal, Eq. (9.22) gives

$$\frac{D_A^{0.5}}{m} \cdot \frac{L_A}{u_A^{0.5} d_{p,A}^{1.5}} = \frac{D_x^{0.5}}{m} \cdot \frac{L_x}{u_x^{0.5} d_{p,x}^{1.5}} \qquad (9.26)$$

where subscripts A and x denote analytical and process scales, respectively. At the same temperature for both scales, Eq. (9.26) reduces to

$$\frac{L_x}{u_x^{0.5}} = \left(\frac{L_A}{u_A^{0.5}}\right)\left(\frac{d_{p,x}}{d_{p,A}}\right)^{1.5} \qquad (9.27)$$

For this particular example, $L_x/u_x^{0.5} = (42.4)(8) = 339.2$. Hence, at the same linear velocity as used on the analytical scale (i.e., $u_x = u_A = 2$ cm/min), $L_x = 480$ cm. The plate height $[H = L/N$, see Eq. (9.4)] is thus estimated to be increased by a factor of 8 when the particle size is increased by a factor of 4. If a 300-μm particle (instead of 200 μm) were to be used on a large scale, the plate height would be increased by a factor of 14.7 over the analytical scale case.

Other variations can be (estimated) from Eq. (9.27), including the effect of reducing eluant linear velocity. Operating temperature may also become a factor, since raising the temperature would decrease L_x by increasing D [see Eq. (9.17)] due to viscosity as well as direct temperature effects on the diffusion coefficient. Calculation of under nonideal conditions (i.e., skewed peaks) is more difficult and is addressed later in this chapter.

The guidelines summarized in this section should be used with caution given constraining conditions inherent in the semiempirical equations presented. Nevertheless, experience shows these relationships to be helpful in obtaining a first estimate of column size and throughput upon scale-up.

2.8. Productivity

The productivity, P, of a column for each cycle depends on the acceptable product purity. As an example, let us consider separation of a two-component mixture containing equal parts of component 1 and 2 at a concentration of X_s (weight fraction) of each in a total sample volume of V_s. A product of specified purity of component 2 is desired. In this case, the volume of product, $V_{p,2}$, obtained having the desired purity has an average concentration of $X_{p,2}$ (note, typically $X_{p,2} < X_{s,2}$). The yield, Y, is then

$$Y = \frac{X_{p,2}V_{p,2}}{X_{s,2}V_s} \tag{9.28}$$

The weight of support, W_s (actual wet weight), is known for this column. The productivity, P, for this cycle is

$$P = \frac{X_{p,2}V_{p,2}}{W_s} \tag{9.29}$$

If both components constitute a desirable product, the productivity would, of course, be higher:

$$P = \frac{X_{p,2}V_{p,2} + X_{p,1}V_{p,1}}{W_s} \tag{9.30}$$

The product fractions that are not of the desired purity must be reprocessed if they are to be recovered at the desired purity. The amount of product recovered for each cycle is determined by experiment.

If the productivity does not change during the operational life of the support, the average productivity $\bar{P}$, is the same as P. If productivity decreases as the support ages, the average productivity, $\bar{P}$, is $\bar{P} = P_{tot}/L$; where P_{tot} is the total weight of product obtained over L number of cycles. The relationship of productivity to resin cost is described later in this chapter.

2.9. Deviations from Ideal Conditions

The preceding description is based on a system in which the components elute in the form of gaussian peaks and in which the sample volume is a small fraction of the overall total column volume. In cases of practical interest, the components being separated may elute as skewed (nongaussian) peaks and the sample volume may, in fact, occupy 10–40% of the column void fraction. The engineering fundamentals for such cases are not as well developed as for the ideal case and would seem to deserve further attention. An empirical approach, however, can still be used to carry out a preliminary analysis of column performance. The second central moment of a chromatographic peak, μ_2, is the variance, σ^2 (Kucera, 1965). Let μ_k be the kth central moment of a function $c(t)$ defined by (Kucera, 1965):

$$\mu_k = \frac{1}{m_o} \int_o^\infty (t - \mu_1')^k \, C(t) \, dt \tag{9.31}$$

where the kth moment of function $C(t)$ is given by

$$\mu_k' = \frac{m_k}{m_o} \tag{9.32}$$

and

$$m_k = \int_o^\infty t^k \, C(t) \, dt \tag{9.33}$$

The parameters for time, t, and the concentration, $C(t)$, represent the ordinate and abcissa, respectively, of a chromatographic peak for a single component under isocratic (constant flow) conditions. Integration of the area under the chromatographic peak gives

$$m_o = \int_o^\infty C(t) \, dt \tag{9.34}$$

The first moment, μ_1', of the curve is

$$\mu_1' = \frac{\int_o^\infty t \, C(t) \, dt}{m_o} \tag{9.35}$$

Numerical values of t and $C(t)$ are determined experimentally. Numerical integration using Eqs. (9.31), (9.34), and (9.35) [with $k = 2$ in Eq. (9.31)] gives the value of σ^2. The plate count can then be estimated using Eq. (9.2).

The effect of large sample volumes on the shape of the peak has been presented by Barford et al. (1978). They demonstrated that the observed resolution, R_s', is a function of sample volume as given by

$$\frac{1}{R_s'} = \frac{1}{R_s} + \left(\frac{v}{\Delta V_R} \right) \tag{9.36}$$

where R_s is the resolution obtained for a small sample volume (i.e., $v \to 0$), v is the sample volume, and

$$\Delta V_R = V_{R,2} - V_{R,1} \qquad (9.37)$$

where $V_{R,2}$ is the observed retention volume for component 2, and $V_{R,1}$ is the retention volume for the first component. Hence, if a separation system is characterized with respect to an analytical scale (small sample volume application), a first estimate of sample loading or resolution can be obtained.

2.10. Preliminary Cost Estimate

A major factor in LC separations is the cost of the support or adsorbent. This cost relative to the quantity of product obtained is given by Ladisch et al., (1984b):

$$C_S = \frac{S}{\bar{P}\eta t} \qquad (9.38)$$

where

C_S = product cost due to support ($/kg product)
S = direct cost of support ($/kg support)
$\bar{P}$ = average productivity $\left(\dfrac{\text{kg product}}{\text{kg support} \cdot \text{cycle}}\right)$
η = turnaround time (cycles/hour)
t = operational life of the support (hours) = L/η, *not* including time during storage or regeneration
L = support life in cycles

One cycle refers to the interval between when the sample is injected and the last eluant is collected immediately before injecting the next sample. The turnaround time, η, refers to the fraction of the cycle (or number of cycles) completed in 1 hr. The productivity, $\bar{P}$, is expressed in terms of the dry weight of product obtained at a specified purity. The cost of the support refers to the actual weight as supplied at a cost S. The operational life is the time elapsed under use (rather than storage) conditions before the support is discarded.

The product obtained per kilogram support is given by $\bar{P}\eta t$. However, $\bar{P}$ will be a function of the manner in which the support loses its operational stability or capacity. If the loss is catastrophic after a certain time, t_1, then

$$\bar{P} = P(l) = \text{constant at } t < t_1$$
$$\bar{P} = P(l) = 0 \text{ at } t > t_1 \qquad (9.39)$$

Consequently, $\bar{P}$ is the same for each cycle, l. If the loss is a first-order decay process approximated by

$$P(t) = P_o \exp(- l/\tau) \tag{9.40}$$

then the total amount of product obtained over a number of cycles, L, is

$$P_{tot} = P_o \int_o^L e^{-l/\tau} \, dl = P_o\tau \, (1 - e^{-L/\tau}) \tag{9.41}$$

where τ is the "time" constant (in cycles) for the loss in productivity. The value of τ^{-1} is the slope of the line obtained by plotting $\ln [P(t)/P_o]$ versus l, while treating l as a continuous parameter. This approach is approximate since the loss in productivity is expressed in terms of a discrete variable (i.e., l), rather than a continuous variable (i.e., t). This reflects the fact that in a chromatographic process the loss in productivity would b measured for each cycle rather than continuously. Based on Eq. (9.41), the average productivity, in this case, is

$$\bar{P} = \frac{P_{tot}}{L} \tag{9.42}$$

If the support is periodically regenerated, the average productivity becomes

$$\bar{P} = \frac{P_{tot,R}}{L_R} \tag{9.43}$$

where L_R is the number of cycles between regenerations, and $P_{tot,R}$ is the total product obtained between regenerations. The productivity of the support is assumed to return to P_o after each regeneration.

The cost, C_S, is asymptotic with respect to $\bar{P}$ (Fig. 9.4a). The relationship in Eq. (9.38) can be expressed linearly by plotting $\bar{P}^{-1}$ versus C_S (Fig. 9.4b). The parameter $\bar{P}^{-1}$ represents the amount of support required for a certain level of productivity. The quantity $S/\eta t$ reflects the support cost per cycle.

If a support costing \$10/kg and having a 2000-hr operation life is used in a system in which a sample is injected every 2.5 hr ($\eta = 0.4$) with $\bar{P} = 0.02$ kg product $\cdot$ (kg support)$^{-1}$ $\cdot$ (cycle)$^{-1}$, the support cost is 1.25 cents $\cdot$ (kg support)$^{-1}$ $\cdot$ (cycle)$^{-1}$ and the product cost, C_S, is \$0.625/kg product. At $t = 8000$ hr, this becomes \$0.156/kg product. At $t = 2000$ hr, $S = \$500/$kg support, and $\bar{P} = 0.02$ at $\eta = 0.4$, the cost C_S is \$31.25/kg product. These examples show how Eq. (9.28) and Figure 9.4 might be useful in making a first estimate of the impact of support cost, capacity, and operational stability on product cost.

The cost of regenerating a support is given by

$$C_R = \frac{R}{\bar{P}} \left(\frac{L}{L_R} \right) \tag{9.44}$$

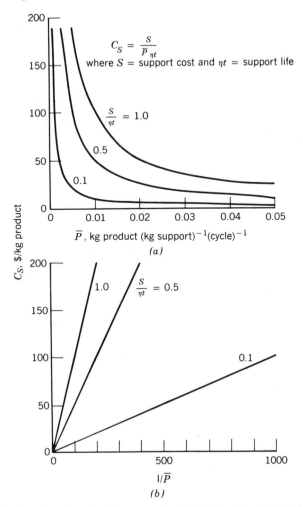

Figure 9.4. Cost functions for estimating support cost: (a) plot of Eq. (9.37) and (b) inverse plot of Eq. (9.37).

where R = regenerant cost in dollars/[(kilogram support)(cycle)] and the other parameters are as defined previously. The ratio L/L_R represents the number of regenerations carried out at regular intervals over the life of the support.

The eluant cost is given by

$$C_E = \frac{M}{\overline{P}} \tag{9.45}$$

where M = solvent cost in dollars/[(kilogram support)(cycle)].

The combined cost is then

$$C_{\text{tot}} = C_S + C_R + C_E \tag{9.46}$$

This should be useful in obtaining a first estimate of a separation cost once a product of satisfactory purity has been attained.

3. SUMMARY

The scale-up of an LC separation requires significant bench and pilot scale developmental effort even if an excellent analytical scale separation is already defined. Equations have been presented that can be used to estimate size, cost, and productivity of a large-scale system based on analytical scale results. Proper definition of operational variables is a key factor in a technically successful scale-up effort. These variables include eluant flowrate, particle size, sample loading, and appropriate column packing procedures. The main objective, in all cases, is to minimize dispersion for a given set of conditions. Although technical feasibility is a necessary condition for commercial applicability, economic feasibility is the sufficient condition. Methods for obtaining a first estimate of separation costs are also described. The combination of technical and economic analyses presented here allows a first estimate of LC scale-up specifications to be made.

ACKNOWLEDGMENTS

Helpful discussions with Dr. Marcio Voloch, Celanese Corporation, Summit, New Jersey made an important contribution in formulating the concepts presented here. A significant portion of this chapter is based on prior work by Voloch and Jacobson as indicated in the citations.

REFERENCES

Absolom, D. R. *Separation and Purification Methods,* **10**(2), 239, 1981.

Antrim, R. L., Collilla, W., and Schnyder, B. J. *Appl. Biochem. Bioeng.,* **2,** 97, 1979.

Anonymous. Design, Operation and Performance of an Industrial Scale HPLC System—Preliminary Technical Notes, Series 300 LC, Elf Aquitaine Development, 9 West 57th Street, New York, NY, 1983.

Anonymous. Biorad. *Chromatography, Electrophoresis, Immunochemistry, HPLC,* Price Richmond, CA, January 1984.

Barford, R. A., McGraw, R., and Rothbart, H. L. *J. Chromatog.,* **166,** 365, 1978.

Beadling, L. C., L. Haff, R. Easterday, and J. Richey, "Chromafocusing: A New Method for High Resolution Protein Purification," Paper 82-59, 184th Am. Chem. Soc. Mtg., MBT Division, Kansas City, Mo, 1982.

Belter, P. A. Recovery Processes—Past, Present and Future, 82-54, 184th Am. Chem. Society Mtg., MBT Division, Kansas City, MO, 1982.

Bier, M., Palusinski, O. A., Mosher, R. A., and SaVille, D. A. *Science,* **219**, 45900, 1281, 1983.

Bird, R. B., Stewart, W. E., and Lightfoot, E. N. *Transport Phenomena,* New York, Wiley, 1960.

Cohen, Y., and Metzner, A. B. *AIChE J., 27*(5), 705, 1981.

DiCesare, J. L., Dong, M. W., and Ettre, L. S. *Chromatoraphia, 14*(5), 257, 1981.

Gareil, P., Durieux, D., and Rosset, R. *Separation Science and Technology, 18*(5), 441, 1983.

Garg, D. R., and Ausikaitis, J. P. *Chem. Engr. Progress, 79*(4), 60, 1983.

Ghin, Y. S., and Chang, H.-N. *Ind. Eng. Chem. Fundam., 21,* 369, 1982.

Hong, J., Voloch, M., Ladisch, M. R., and Tsao, G. T. *Biotechnol. Bioeng., 24,* 725, 1982.

Jacobson, B. J., *M.S. Thesis,* Purdue University, Department of Agricultural Engineering, December 1982.

Katzen, R., Ackley, W. R., Moon, G. D., Messick, J. R., Burch, B. F., and Kaupisch, K. F. *Low Energy Distillation Systems,* 180th Natl. ACS Mtg., Las Vegas, NE, 1980.

Knox, J. H. *Chromatogr. Newsletter, 2*(1), 1, 1973.

Kucera, E. *J. Chromatog., 19,* 237, 1965.

Ladisch, M. R., and Dyck, K. *Science, 205*(4409), 898, 1979.

Ladisch, M. R., Voloch, M., Hong, J., Bienkowski, P., and Tsao, G. T. *Ind. Eng. Chem. Proc. Des. Dev.* **23,** 437, 1984a.

Ladisch, M. R., Voloch, M., and Jacobson, B. *Biotechnol. Bioeng. Symp. No. 14,* 525, 1984b.

Ladisch, M. R., and Tsao, G. T. Vapor Phase Dehydration of Aqueous Alcohol Mixtures, U.S. Patent 4,345,973, August 24, 1982.

Martin, R. L. M., and Synge, A. J. P. *Biochem. J., 35,* 1358, 1941.

Michael, A. L. *Chem. Tech., 35,* 36, 1981.

Pieri, G., Piccardi, P., Muratori, G., and Luciano, C. *La Chimica E L'Industria,* **65**(5), 331, 1983.

Rebar, V., Fishback, E. R., Apostolopoulos, D., and Kokini, J. F. *Biotechnol. Bioeng., 26*(5), 513, 1984.

Regnier, F. E. *Science, 222*(4621), 245, 1983.

Snyder, L. R., and Kirkland, J. J. *Introduction to Modern Chromatography,* New York, Wiley, 1979, 2nd ed.

Voloch, M., Hong, J., and Ladisch, M. R. Dehydration of Ethanol Using Cornmeal as an Adsorbent, MICR Div., 180th Nat'l. ACS Mtg., Las Vegas, NE, 1980.

Wankat, P. C. Operational Techniques for Adsorption and Ion Exchange, presented to the Corn Refiner's Association Conference, June 1982.

10

MEMBRANE SEPARATION TECHNOLOGY: AN OVERVIEW

GEORGES BELFORT

Professor of Chemical Engineering
Rensselaer Polytechnic Institute
Troy, New York

1. INTRODUCTION

Since Abbé Nollet in 1748 discovered the phenomenon of osmosis with a semipermeable animal bladder (membrane) separating wine and water, understanding the transport phenomena and later developing synthetic membranes with the desired properties have been intriguing problems (Ferry, 1936). Fick (1865) in 1865 formulated the first laws of diffusion through colloidon membranes for solutions, while Graham (1866) (1860–1861) discussed selective gas diffusion and the process of dialysis. Zsigmondy (1918) (1907–1908) developed colloidon microporous membranes in Europe, while Donnan published his distribution law based on potential differences. Teorell (1935) and Meyer and Sievers (1936) conceived the electrodialysis process, and Kolf and Berk in 1944 developed dialysis for the artificial kidney.

The Sartorius Werke GmbH in Göttingen, Germany, had begun commercial membrane manufacture as early as 1929 with microfilters made from cellulose nitrate and small dialysis membranes from cellophane. Immediately after World War II, Millipore Corporation was established in the United States mainly to develop membranes for bacteriological analysis (Lonsdale, 1984). At the same time, the electrodialysis process was developed at Toagepast Natuurwetenschappelijk Onderzoek, The Netherlands in the Netherlands and at Ionics in the United States. Paper-based cellulosic membranes were used in one of the first electrodialysis pilot plants built by the Council for Scientific and Industrial Research, South Africa, and later a 100,000-gal/hr full-size plant to treat brackish mine waters from the Free State gold mines in South Africa was designed but never built (Wilson, 1960). Meanwhile, synthetic membranes were increasingly being used for bacterial assays (see Standard Methods for the Examination of Water and Wastewater, 1980).

Although Breton and Reid (1959) in the late 1950s are credited for initiating the idea of a salt-rejecting membrane with preferential water flux, Loeb and Sourirajan (1982) are responsible for making the first asymmetric cellulose acetate membrane with high relative solvent permeation rates. Soon thereafter several industrial groups realized the commercial relevance of this advance and developed elegant ways of housing these asymmetric membranes. For example, Mahon (1963) at Dow Chemical Company spun cellulose triacetate hollow fiber membranes, and Mahon et al. (1969) prepared permselective aromatic polyamide hollow fibers. In 1968, Westmoreland and later Bray developed the successful jelly-roll or spiral-wound module. Although plate-and-frame modules have been developed for reverse osmosis, the earliest versions were too cumbersome, and only relatively recently has this design been offered commercially for reverse osmosis (DDS Corporation, Nakskov, Denmark).

Besides these developments, new composite membranes have been made by layering the salt-rejecting film on top of a porous support film (Cadotte and Petersen, 1980; Kurihara et al., 1981). Thus, both films could be optimized for performance without compromising one for the other.

As an outgrowth of hyperfiltration (or reverse osmosis), a relatively recent low-pressure membrane process, ultrafiltration (UF), has become commercial (Strathmann, 1984). Although the concept of UF (and microfiltration, MF) was developed during the first half of this century, its commercial and technical significance was not fully recognized until later in the United States. Recent developments in casting microfilters in tubular or capillary geometry have been reported in Europe, Japan, the United States, and Australia. These latter more open filters (> 0.1-μm pore size) are finding application in the biotechnology, chemical, and wastewater treatment industries. Besides hydrophobic polypropylene, other materials such as polysulphone and polyamide are being used to form these membranes.

Several groups attempted to describe the mass transport toward and through membranes after the 1960s (Kleinstreuer and Belfort, 1984). The most prominent of these has been Staverman (1951), Spiegler and Kedem (1966), and Pusch and Woermann (1970) who have emphasized the thermodynamics of irreversible (IR) processes, whereas others such as Sherwood et al. (1965) published the first analyses on the concentration polarization of solute during the selective transport process. Others, as described in Merten's book (1966), formulated the solution–diffusion transport model in an attempt to describe transport of solute and solvent and thus assist in developing new membranes. Kamiyama et al. (1984) presented an extremely useful correlation between water flux and NaCl rejection showing the regions for several different polymer membranes. This is shown in Figure 10.1.

The single most limiting factor in the performance of hyperfiltration (HF), ultrafiltration, microfiltration, and electrodialysis is concentration polarization (Jonsson and Boessen, 1984). For the first three processes, transmembrane flux is adversely effected by the transient build up of retained solutes (salt, colloids, macromolecules) and resultant presence of high solute concentrations at the upstream solution–membrane interface. For electrodialy-

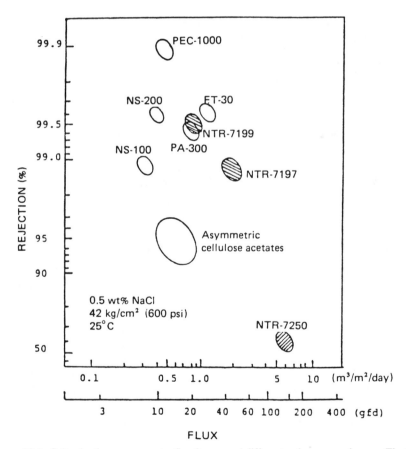

Figure 10.1. Salt rejection versus water flux for several different polymer membranes. The FT-30 is a polyamide type composite membrane. (Kamiyama et al., 1984.)

sis, a similar phenomenon occurs, except that the solute (the major transferred species) becomes depleted at the permselective membrane–dialysate
interface, resulting in water splitting and concomitant ohmic heating and
process inefficiencies. To reduce concentration polarization, modules were
developed to allow the feed streams to flow tangentially along the membrane
surface. By using hydrodynamic considerations, polarized solutes could be
sheared from the membrane surface, thereby increasing the back diffusion
and reducing the decline in performance. At first, it was thought that turbulent flow conditions were most desirable, but later, when energy considerations became paramount, laminar flow in ultrathin channels was used for
superior polarization control. An example is the development of commercially available ultrafiltration hollow fibers with high-membrane area:volume
ratio and high volumetric throughput. In fact, reverse depolarization (back
flushing) is used to clean the surface of these membranes.

Because of the seminal role that fluid mechanics plays in reducing concentration polarization in pressure-driven (and electrically driven) membrane
processes, it will be discussed in depth here. To do so, however, some
background in membranes, modules, and basic fluid–particle mechanics in
porous ducts is needed and will be presented in Sections 2–4. These subjects
will be followed by a detailed discussion in Section 5 on concentration
polarization [with reference to Matthiasson and Sivik's paper (1980)]. After
that, in Section 6 details on fouling in pressure-driven membrane processes
will be presented. Finally, current uses, limitations, and future developments on the horizon will be presented.

2. MEMBRANES

2.1. Criteria

In general, a commercially attractive membrane should be (1) semipermeable, (2) high enough in rates of transport to be economic, (3) chemically and
mechanically able to withstand excursions in pH and temperature, different
solvents, and relatively high pressures, (4) nonbiodegradable, and (5) easy
and economical to cast.

2.2. Classification

The different membrane processes can be classified according to their driving force or flux and are usually expressed as a linear analog of Ohm's or
Fourier's law (Table 10.1); that is, they only account for the diagonal
coefficients in the IR theory discussed in Section 4. All the off-diagonal
coefficients are considered negligible.

Membranes are also classified according to their final morphology and
type (Table 10.2) or method of formation. Pusch and Walch (1982) divide

TABLE 10.1. Classification of Membrane Processes[a]

Law	Expression	Flux	Force	Resistance	Process
Fick	$V_w = -D\Delta C$	V_w	$\Delta C(\Delta\pi)$	D^{-1}	Dialysis
Ohm	$I = R^{-1}E$	I	E	R	Electrodialysis
Darcy	$V_w = L_p(\Delta P - \sigma\Delta\pi)$	V_w	$(\Delta P - \sigma\Delta\pi)$	L_p^{-1}	Reverse osmosis
(Hagen-					Ultrafiltration
Poiseuille)					Microfiltration
					Piezodialysis

[a]The constants (resistance) for the linear laws are assumed constant with respect to flux and force but could be a function of temperature, mole fraction, and so on.

membranes according to three mechanisms of separation: sieving, electrostatic, relative solubility (partition) effect. They divide membrane type into three categories as shown in Figure 10.2: coarsely porous, finely porous, and nonporous, assuming that solute–membrane interactions are negligible. Their relative pore sizes are greater than 50 Å (e.g., porous glass frit); between 10 and 50 Å (e.g., UF membranes); and < 10 Å (oil film or HF or reverse osmosis). In Figure 10.2, σ is the Staverman reflection coefficient and varies from 0 (no solute rejection) to 1 (complete rejection).

2.3. New Materials

Choosing a new potentially interesting membrane material has largely been an art form, with a search for polymers possessing the ability to dissolve in a solvent allowing the membrane to be cast via a phase-inversion process. Organic chemists were then able to choose a variety of polar groups [= O, —OH, —SO$_4$, —N] and apolar groups [—CH$_3$, —OCH$_3$] and to attach them in various ways to aromatic or aliphatic aromatic polymers.

The newly formed synthetic membrane is usually tested and compared

TABLE 10.2. Membrane Morphology and Type[a]

Type	Pore Size Distribution	Examples	Process[b]
Woven	Wide and large ($> \mu m$)	Screen, cloth	F
Sintered (Moldered)	Normal and narrow ($> \mu m$)	Metals, ceramics	MF
Cast and extruded	Narrow and wide ($> nm$) (Symmetric and asymmetric)	Polymer solution	MF, UF, HF
Irradiated	Very narrow (>30 nm)	Polycarbonate	MF

[a]Membrane structures could also be classified by their pore geometry, that is, cylindrical, modular, spongelike, and nonporous.
[b]Symbols are F, filtration; MF, microfiltration; UF, ultrafiltration; and HF, hyperfiltration or reverse osmosis.

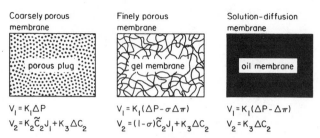

Figure 10.2. Models of different homogeneous synthetic membranes. $K_2 = C_2/c_2$ is the solute partition coefficient, C_2 the solute concentration within the membrane at the surface in contact with the external solution, c_2 the solute concentration of the external solution, $\bar{c}_2$ an average solute concentration that can be calculated from the external concentrations c_2' and c_2'' at both sides of the membrane, $P_2 = K_2 D_{2_m}$ the solute permeability, D_{2_m} the concentration-independent solute diffusion within the membrane, d the effective thickness of the membrane, $\Delta c_2 = c_2' - c_2''$ the solute concentration difference across the membrane (driving force of the diffusive part of the solute flux), and V_1 the volume flux, L_p = hydrodynamic permeability, σ = Staverman reflection coefficient. (Pusch and Walch, 1982.)

with the best performing currently available membrane (material). The chemical and physical structures of membranes are studied using many methods (Table 10.3). Of these methods, the most widely used, besides direct permeation studies, is electronmicroscopy. Hansmann and Pietsch (1949) and later Helmcke (1954) were the first to obtain electron microscopic photographs of membranes and to rank them according to pore size. Later Frommer and Lancet (1972), using elegant but simple light microscopic methods, provided unusual insight into the membrane-formation mechanism.

2.4. Membrane Formation

As mentioned above, after Loeb and Sourirajan (1982) showed that membrane thickness (i.e., $\sim$ 2000-Å dense skin) was inversely related to permeation flux, different materials besides cellulose acetate were cast into successful membranes. Following this, two groups prepared membranes with individually optimized skin and understructure. Rozelle et al. (1973) and Riley et al. (1971) succeeded in fabricating such "composite" membrane structures in the early 1970s (Fig. 10.3).

During membrane formation both thermodynamic (phase separation) and transport (diffusion phenomena need to be considered. After formation, heat treatment (annealing) can be used to change the membrane permeation characteristics. Typical thermodynamic processes that occur during membrane formation are summarized in Table 10.4.

Strathmann et al. (1975) and more recently Bokhorst et al. (1981) have used the ternary phase diagram to identify the path taken by the polymer solution in forming the skin and understructure of reverse osmosis mem-

TABLE 10.3. Methods for the Characterization of the Structure of Synthetic Membranes

Microstructure (pore diameter <50 Å)	Macrostructure (pore diameter >50 Å)
Diffraction of slow neutrons	Electron microscopy
Gas permeation with the help of the free-volume theory	Gas and liquid permeation using the Hagen-Poiseuille law and the Knudsen relationship
Gas adsorption using the BET isotherm, the dual-sorption model, and the free-volume theory	Porosimetry (e.g., Hg-intrusion method)
Thermodynamic characterization of the water structure	Bubble pressure method
Thermomechanical analysis (TMA)	
Determination of the limits of molecular separation	
IR absorption spectroscopy	
X-Ray scattering and diffraction	X-Ray small-angle scattering
Differential thermoanalysis and scanning calorimetry (DTA and DSC)	

Source: Pusch and Walch, 1982.
BET, Brunauer Emmett and Teller; DTA, differential thermal analysis; DSC, differential scanning calorimeter.

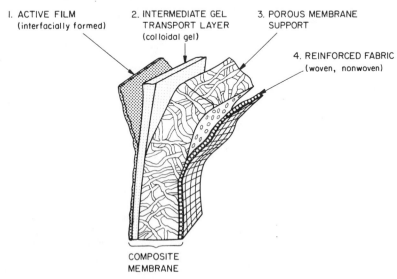

1. ACTIVE FILM (interfacially formed)   2. INTERMEDIATE GEL TRANSPORT LAYER (colloidal gel)   3. POROUS MEMBRANE SUPPORT   4. REINFORCED FABRIC (woven, nonwoven)

COMPOSITE MEMBRANE

Figure 10.3. Cross section of a typical thin film composite membrane.

TABLE 10.4. Thermodynamic Considerations

Process	Characteristics
Liquid–liquid phase separation	Nucleation and growth spinodal decomposition
Crystalization	Free energy of fusion slow process
Gelatin	Polymer–polymer contact (cross linkage)

branes (Fig. 10.4). Also shown in Figure 10.5 are the transport processes occurring during asymmetric membrane formation. After its formation the asymmetric membrane is immersed into a hot water bath. The flux and rejection have the typical relationship shown in Figure 10.6. The annealing process results in membrane shrinkage and densification of the skin.

2.5. Scanning Electron Micrographs

Scanning electron micrographs (SEM) of several asymmetric and symmetric membranes are shown in Figures 10.7 and 10.8, respectively. Two hollow fiber configurations of a polysulfone and a regenerated cellulose acetate membrane are shown in Figure 10.7a and b. The skins are on the bore side of the fiber, and the spongy support matrix extends radially outward. Large open voids are clearly visible. A flat polyamide asymmetric membrane and an RC100 composite membrane are shown in Figure 10.7c and d, respectively.

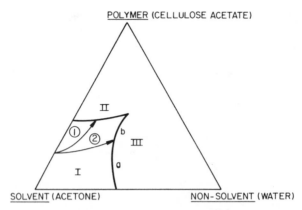

Figure 10.4. Ternary phase diagram for polymer (cellulose acetate), solvent (acetone), and nonsolvent (water). Three regions are identified as homologous solution (I), gelation region with high polymer concentration (II), and liquid–liquid phase separation region with separation due to nucleation and growth (IIIa) and spinodal decomposition (IIIb). During membrane formation, skin formation (path 1) and porous sublayer formation (path 2) occur simultaneously. (Bokhorst et al. 1981).

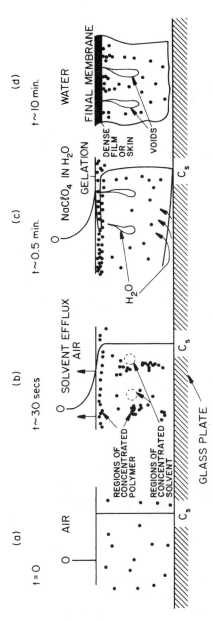

Figure 10.5. Transport processes during asymmetric membrane formation on a glass plate in air. (*a*) Initial solvent concentration profile; (*b*) after solvent evaporation begins, polymer concentration increases at the air–solution interface; (*c*) further loss of solvent near the interface results in polymer gelation during immersion in cold (4°C) aqueous NaClO$_4$ solution. Nonsolvent (water) enters membrane matrix dissolving solvent pockets and forming voids or fingers (*d*). Final membrane formed after annealing in water for about 10–30 min at 75 < *t* < 95°C.

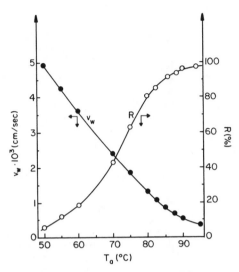

Figure 10.6. Volume flux and salt rejection R of paper-reinforced asymmetric flat sheet CA membranes (KALLE Niederlassung de Hoechst AG) as a function of the annealing temperature T_a (brine = 0.1 mol/liter NaCl; ΔP = 100 atm; t = 25°C). (Pusch and Walch, 1982.)

Several symmetric membranes are shown in Figure 10.8. These include a hollow fiber porous glass membrane and a stretched polypropylene membrane, shown in Figures 10.8a and b, respectively. In Figures 10.8c and d we see the side and plan views of a polycarbonate sheet membrane, respectively.

2.6. Performance

The flux and retention characteristics for many membrane materials (and derivatives thereof) are summarized for hyperfiltration in Table 10.5a (cellulose acetate) and Table 10.5b (noncellulosic) and for ultrafiltration in Table 10.6.

3. MEMBRANE PERMEATORS OR MODULES (Belfort, 1984)

The main requirement of a membrane permeator or module is that it house the membranes in such a way that the feed stream is sealed from the permeate stream. Other requirements are concerned with the following:

1. Mechanical stability, such as supporting a fragile membrane under very high differential pressures (200–1500 psi) for ultrafiltration and hyperfiltration and under high electrical fields for electrodialysis, preventing pressure leaks between feed and product streams and between

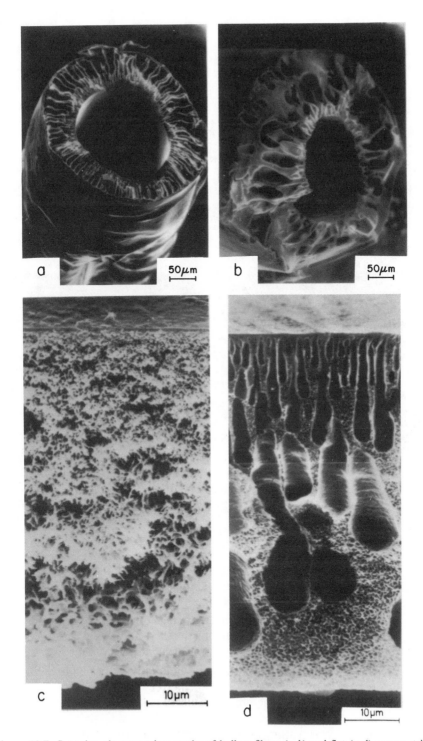

Figure 10.7. Scanning electron micrographs of hollow fibers (*a,b*) and flat (*c,d*) asymmetric membranes. The membranes are (*a*) polysulfone, (*b*) regenerated cellulose acetate, (*c*) poly-amide, and (*d*) RC100 composite. (Pusch and Walch, 1982.)

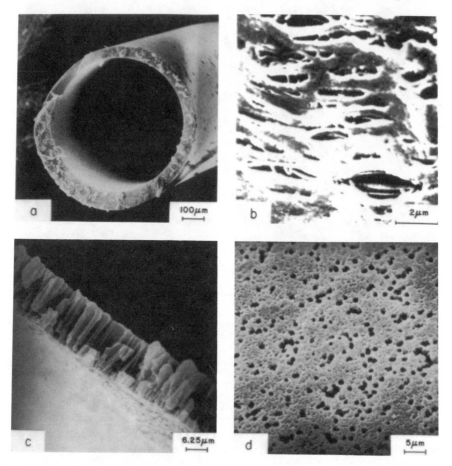

Figure 10.8. Scanning electron micrographs of symmetric membranes. The membranes are (*a*) porous glass, (*b*) stretched polypropylene sheet (Pusch and Walch, 1982), and the side (*c*) and plan (*d*) views of a polycarbonate sheet membrane.

the feed stream and its surroundings (air) and avoiding large pressure drops in the feed and product streams

2. Hydrodynamic considerations, such as minimizing concentrated polarization, including the build up of solute and fouling layers on the membrane surface, to impede membrane performance

3. Economic considerations, such as obtaining high membrane-packing density to reduce capital costs of the pressure vessels and designing the unit for ease of membrane replacement.

Commercial permeators that meet these requirements can be classified into five broad design categories based on the membrane geometry: tubular,

TABLE 10.5a. Commercial Hyperfiltration Membranes and Modules from Cellulose Derivatives

Membrane Material[a] (Cellulose Derivative)	Module[b]	Manufacturer	R [%]	Q [$L\,m^{-2}\,d^{-1}$]	Test Conditions
-2.5-Acetate	T, in situ, asymm.	Abcor	<96	>400	40 bar, 1% NaCl, pH 3–7, 35°C
-2.5-Acetate	S, asym.	Babcock (Ajax Int.)	95	700	40 bar, 0.5% NaCl, pH 4–6, 30°C
-Acetate, quaternated	S, asym., dry	Chem. Systems	99	500	68 bar, 1% NaCl, pH 4–8, 32°C, 0.5 ppm Cl$_2$
-Acetate	S	Daicel			—
-2.5-Acetate	F, asym.	DDS	<99	>400	40 bar, 1% NaCl, pH 2–8, 30°C
-3-Acetate	H, sym., melt extraction	Dow Chem.	98.7	30	56 bar, 3.5% NaCl, pH 4–7.5, 35°C, 0.5 ppm Cl$_2$
-3-Acetate	H, asym.	Dow Chem.	97	200	28 bar, 0.3% NaCl, pH 6–8, 35°C
-Acetate, mixture	S, asym.	Envirogenics	98	1000	42 bar, 0.2% NaCl
-Acetate, mixture	F, asym.		97.5	600	80 bar, 3.5% NaCl
-Acetate-butyrate	asym.	Envirogenics	99.8	400	100 bar, 3.5% NaCl
Acetate, mixture	S, asym., dry	Hydranautics	95–97	1000–600	28 bar, 0.1% NaCl
-2.5-Acetate (Nadir)	T, F, asym., dry	Kalle (only membranes)	<98	>400	40 bar, 0.5% NaCl, pH 3–7.5, 30°C
-2.5-Acetate	S, asym.	Millipore	90		14 bar, 0.1% NaCl
-Acetate, mixture	T, asym.	Nitto Electric	93		50 bar, 3.5% NaCl
-Acetate	F, W	Osmonics	<99	>400	56 bar, pH 2–3, 45°C
-2.5-Acetate	T, in situ, asym.	Paterson Candy	<98	>500	40 bar, 0.5% NaCl, pH 3–6, 30°C
-3-Acetate	F, asym.	Sartorius	99.8	<300	105 bar, 3.5% NaCl
-Acetate	F, asym.	Schleicher & Schüll	<99	>400	80 bar, 3.5% NaCl, pH 4–7, 40°C
-Acetate	S[c]	Toray Ind.	96	300	56 bar, 3.5% NaCl, pH 4–7, 25°C, 0.5 ppm Cl$_2$
-2-Acetate	H	Toyobo Co.	93	—	30 bar, 0.2% NaCl, pH 3–7, 30°C, 0.5 ppm Cl$_2$
-Acetate	T	UOP	97	450	40 bar, 0.5% NaCl, pH 3–7, 45°C
-2.8-Acetate	S, composite	UOP	>99	400	70 bar, 3.5% NaCl, pH 4–7.5, 35°C, 0.5 ppm Cl$_2$
-2.5-Acetate	T, asym.	Wafilin	<96	>700	40 bar, 0.5% NaCl, pH 3–8, 35°C
-Acetate	T	Western Dynetics	90	700	35 bar, 0.5% NaCl, pH 3–7, 40°C

Source: Pusch and Walch, 1982.

[a] The number preceding "acetate" signifies the average number of acetylated OH-groups per glucose unit.

[b] Abbreviations: H, hollow fiber; R, salt rejection; Q, volume flux.

[c] Tangential flow.

251

TABLE 10.5b. Hyperfiltration Membranes and Modules from Noncellulosic Material

Membrane Material[a]	Module[b]	Manufacturer	R [%]	Q [$L\,m^{-2}\,d^{-1}$]	Test Conditions
Polysulfone, sulfonated	H, asym.	Albany	—	—	pH 2–13, 70°C, 1 ppm Cl_2
Polybenzimidazole	H, asym.	Celanese	99.4[d]	50	70 bar, 3.5% NaCl, pH 2–11, 75°C
LP-Polyamide	S, composite	Desalination[c]	96	1000	14 bar, 0.1% NaCl
Polyamide, aromatic, B 9	H, asym.	Du Pont[c]	95	50	28 bar, 0.15% NaCl, pH 4–11, 35°C
Polyamide, aromatic, B 10	H, asym.	Du Pont[c]	98.5	40	56 bar, 3.5% NaCl, pH 5–9, 35°C
Polyamide-/hydrazide	S, asym.	Du Pont	99	300	70 bar, 3.5% NaCl, pH 5–9, 35°C
p-Phenylenediamine/ Trimesoyl chloride, oxidized	F, composite	Film Tec[c]	98	1800	40 bar, 1% NaCl, <60°C, 0.5 ppm Cl_2 (25°C)
Polyfuran, NS-200	H, composite	FRL	99.5	400	56 bar, 0.2% NaCl
Polyphenylene oxide, sulfonated	composite	General Electric	84	1500	77 bar, 0.1% NaCl, pH 2–13, 54°C, 1 ppm Cl_2
Piperazine/trimesoyl chloride/isophthaloyl chloride	composite	MRI	50	1700	14 bar, 0.5% NaCl
Polyethylene imine/ Toluylene diisocyanate, NS-100	composite	NSRI	99.4[e]	900	100 bar, 3.5% NaCl, pH 1–13
Polyfuran NS-200	composite	NSRI	99	1000	70 bar, 3.5% NaCl, pH 2–12, 90°C
ZrO_2/Polyacrylic acid	dynamically shaped	ORNL	90	7000	70 bar, 0.3% NaCl
Polyfuran	S, composite	Osmonics[c]	99	200	56 bar, pH 0.5–10.5, 75°C
Glass, phase separation	H[g]	Schott	99.6	100	100 bar, 0.5% NaCl
Co-Polyacrylonitrile[h]	F, T, asym.	Sumitomo Electric Industries[c]	95–98	1000–3000	50–100 bar, 0.6% NaCl, pH 1–10, 45°C, 0.1 ppm Cl_2
Polybenzimidazolone	asym.	Teijin[c]	99.5[f]	800	80 bar, 1% NaCl
	S, composite	Teijin[c]	97–99.3	1000	100 bar, 0.7% NaCl, pH 1–12, 60°C
Piperazine/Polyepiiododohydrin/ Isophthaloyl chloride	S, composite[i]	Toray[c]	99.7	700	40 bar, 0.25% NaCl, pH 4–12, 40°C
Polyamide TFC-802	composite, PA-100	UOP	99	750	70 bar, 3.5% NaCl
Ethylenediamine/polyepichlor- hydrin/Isophthaloyl chloride	composite, PA-300	UOP	99.4	1200	70 bar, 3.5% NaCl, pH 2–12, 55°C
Polyether-/amide, TFC-803	S, composite	UOP[c]	98.0	600	56 bar, 3.5% NaCl, pH 2–12, 45°C
Polyether-/urea, TFC-801	S, composite, RC-100	UOP[c]	99.2	600	63 bar, 3.5% NaCl, pH 2–12, 45°C

Source: Pusch and Walch, 1982.

[a] The letters and/or numbers after some of the names signify tradenames or type designations.
[b] Abbreviations: H, hollow fiber; Q, volume flux; R, salt rejection.
[c] Commercial product.
[d] R (urea)—88%.
[e] R (phenol)—84%.
[f] R (urea)—72%.
[g] In presence of alkyl sulfonates.
[h] Cold plasma discharge. Solrox process.
[i] Tangential flow.

TABLE 10.6. Commercial Ultrafiltration Membranes and Modules

Membrane Material[a]	Module[b]	Manufacturer	MW Cutoff	Q [$L\ m^{-2}\ d^{-1}$]	Test Conditions
Cellulose 2-acetate (HFA)	S, asym.	Abcor	1500	6000	3.5 bar, 50°C, pH 3–7.5, 1 ppm Cl_2
	T, in situ, asym.			10000	3.5 bar, 90°C, pH 2–9, 10 ppm Cl_2 (25°C, pH 0.5–13)
Polyvinylidene fluoride (HFM)	S, asym.	Abcor	10000/20000	7500	
	T, in situ, asym.			30000	
Polytetrafluorethylene	F, sym.	Aqua-Chem.	50000	4000	5.6 bar, 100°C, pH 1–14
Co-Polyacrylonitrile (HC-1, HC-5, HI-1)	C, sym.	Asahi Kasei / Asahi Chemical Ind.	6000/13000	1000/4000	3 bar, 50°C, pH 2–10
Polybenzoxazindione, sulfonated	F, asym.	Bayer	300	2000	40 bar, 50°C, pH 2–12
Polyamide-/imide (BM)	C, asym.	Berghof/Nucleopore	2000–50000	100–6000	1.5 bar, 80°C, pH 2–12
Co-Polyacrylonitrile/Vinylpyrrolidone	F, asym.	Daicel	30000	2000	3 bar
Cellulose acetate (800–500)	F, asym.	DDS	6000–65000	2000–7000	5 bar, 50°C, pH 2–8, 0.002% NaOCl
Polysulfone (Gr 8/6/5 P, GS 81 PP sulfonated)	F, asym.	DDS	8000/20000/65000	1000–7000	3 bar, 80°C, pH 0–14, 2% NaOCl
Modacryl polymer	P, asym.	Dorr-Oliver	1800/24000	3000/12000	2 bar, 50°C, pH 2–11.5
Polysulfone	P, asym.	Dorr-Oliver	10000	10000	2 bar, 70°C, pH 2–12
Celulose acetate (Nadir)	F, T, asym.	Kalle	2000–100000	400–10000	3 bar, 35°C, pH 2–8, 0.002% NaOCl
Polyamide (Nadir)	F, T	Kalle	20000–100000	2000–10000	3 bar, 65°C, pH 2–12
Polysulfone (Nadir)	F, T, asym.	Kalle (Membranes exclusively)	10000–60000	1000–12000	3 bar, 90°C, pH 1–14, 2% NaOCl
Polysulfone (PTUF)	S, asym.	Millipore	80000	2000	5 bar, 32°C, pH 3–11
Polyolefin, hydrophilic	T, asym.	Nitto Electric	20000/100000		4 bar, 40 °C, pH 2–10
Cellulose acetate	F, W, asym.	Osmonics	1000/20000	1000/6000	3.5 bar, 27°C, pH 2–8
Polysulfone	F, W, asym.	Osmonics	1000/20000	4000/12000	3.5 bar, 93°C, pH 0.5–13
Cellulose acetate (T2-5A)	T, asym.	Paterson Candy	1000–20000		pH 3–6, 30–50°C
Polysulfone (PM)	C, asym.	Romicon/Amicon	2000–50000	1000–14000	1 bar, 75°C, pH 1.5–13
Modacryl polymer (XM/GM)	C, asym.	Romicon/Amicon	50000/80000	12000–15000	1 bar, 50°C, pH 1.5–13
Co-Polyacrylonitrile/Methallyl sulfonate	F (Iris 3042)	Rhône-Poulenc	20000	7000	2 bar, 40°C, pH 1–10
Polysulfone, sulfonated	F, asym. (Iris 3022)	Rhône-Poulenc	20000	15000	2 bar, 80°C, pH 1–13
Carbon/ZrO_2	T, in situ	S.F.E.C.	50000	—	100°C, pH 1–14
Polybenzimidazolone	T, asym.	Teijin	—	—	60°C, pH 1–12
Polysulfone	T, asym.	Wafilin	5000	8000	3 bar, 95°C, pH 1–12
Polyacrylonitrile	T, asym.	Wafilin	10000	7000	3 bar, 60°C, pH 4–10
Cellulose (CUF)	T, asym.	Western Dynetics	20000	2500	1 bar, 70°C
Polysulfone (PSUF)	T, asym.	Western Dynetics	20000	2500	1 bar, 80°C

Source: Pusch and Walch, 1982.

[a] The letters and/or numbers after some names signify tradenames or type designations.

[b] Abbreviations: F, flat sheet; C, capillary membrane; P, plate; T, tube; S, spiral-wound module; Q, volume flux.

253

spiral wrap, hollow fiber, flat plate, and dynamic. Nearly all of these permeators or variants thereof have been produced commercially.

For example, reverse osmosis subclasses within each subcategory are described in Table 10.7. A sketch of each of the major commercial reverse osmosis membrane permeators is shown in Figures 10.9 and 10.10. Several performance and structural characteristics for the different permeators are also presented in Table 10.8.

The first thing to notice in Table 10.8 is that the permeator with the lowest water output per unit volume (tubular with inside flow) is most easily cleaned, whereas the permeator with the highest water output per unit volume (fibers with brine flow on the outside) is the most difficult to clean. Ease of cleaning is especially important for a turbid feed such as wastewater or a typical fermentation broth. Unfortunately, it is the permeators, which have not yet been scaled-up to very large volumes, that offer the best of both worlds: (1) the tubular design with brine flow on the outside, (2) the fiber design with brine flow on the inside, and (3) the dynamic membrane concept. The flow channel size, in the penultimate column in Table 10.8, is presented as a measure of the cross-sectional area available for brine flow. In a crude calculation, a large value of flow channel size should indicate little chance of fluid holdup due to blockage from floatables or suspended solids. A small value should suggest that a high degree of prefiltering is necessary.

3.1. Helical and Rigid Tube Permeators (Figure 10.9)

The membrane is cast as a tube on the inside of a porous support tube (e.g., paper or cloth), which is placed inside a pressure vessel. The brine stream flows through the tube while the product permeates the membrane radially. The pressure vessel may be a steel pipe with perforated holes (rigid design), or, if the support tube can withstand the pressure differential, a plastic or low-pressure housing (helical design) can collect the product. Rods or spheres sometimes placed inside the tube, along the center line, increase fluid velocity and axial shear at the membrane–solution interface. Sanderson (private communication, 1980) reports the production of a relatively inexpensive tubular vessel made from cast plastic elements held together with a rod. By slightly misaligning the elements, the cast membrane tube will have protrusions desirable for fluid mixing. Reynold's numbers as high as 130,000 have been used in tubular systems (Sachs et al., 1975). Tubular units are easily cleaned, and much operating data exist for them. Their disadvantages include low water production per unit volume, high water holdup per unit area of membrane, and relatively expensive membranes (about $10–20/ft^2).

3.2. Spiral-Wrap Permeators (Figure 10.10)

Several flat or planar membranes are sandwiched between porous plastic screen supports and then formed into a "swiss roll." The edges of the

TABLE 10.7. Reverse Osmosis Membrane Permeators

Class	Designation	Description of Available Designs	Manufacturer[a]
Tubular	1a	Brine flow inside straight rigid support tube	A,C,E,L,M, N,P,Q
	1b	Brine flow inside helical support tube	
	1c	Brine flow inside straight squashed tube	Q
	1d	Brine flow outside straight rigid support tube	R
	1e	Brine flow outside flexible rigid support tube	Q
Spiral wrap	2a	Brine flow between alternate leaves of a spiral wrap	C,E,J,P,S,U
Fiber	3a	Brine flow outside flexible hollow-fiber membranes	B,G,I,T
	3b	Brine flow inside flexible hollow-fiber membranes	—
Flat plate	4a	Horizontal filterpress design with brine flow radially between leaves	F,K
	4b	Same as 4a with whole unit spinning	H
Dynamic membrane	5a	A dynamic precoat membrane is laid down on a porous support	D,O

Source: Belfort, 1984.

[a] The following letters are used to designate major manufacturers: A, Abcor, Inc., Wilmington, Massachusetts; B, Asahi Chemical Co., Tokyo, Japan; C, Calgon Havens Systems, San Diego, California; D, Carre, Inc., Seneca, SC; E, Culligan International Co., Northbrook, Illinois; F, DeDanske Sukkerfabrikker, DK-1001, Copenhagen, Denmark; G, Dow Chemical Co., Walnut Creek, California; H, Dresser, Advanced Technology Center, Burroughs, Irvine, California; I, E. I. DuPont de Nemours & Co., Wilmington, Delaware; J, Envirogenics Co., El Monte, California; K, GKSS-Forschungszentrum, Germany; L, Israel Desalination Engineering, Tel Baruch, Israel; M, Kalle, West Germany; N, Nitto Electric Co., Osaka, Japan; O, Oak Ridge National Laboratory, Oak Ridge, Tennessee; P, Osmonics, Inc., Hopkins, Minnesota; Q, Paterson Candy Int., Laverstroke Mills, Whitchurch, Hamps, U.K.; R, Raypak, Inc., Westlake Village, California; S, Toray Industries, Inc., Otsu, Shiga, Japan; T, Toyobo Co., Ltd., Katata Research Center, Otsu, Shiga, Japan; U, UOP Fluid Systems Div., San Diego, California.

membranes are sealed to each other and the central perforated tube. The resultant spiral-wrap module is fitted into a tubular steel pressure vessel, such as a 4-in. nominal pipe. The pressurized feed solution is fed into the pipe so that it flows through the plastic mesh screens along the surface of the membranes. The product, which permeates the membranes, flows into the closed alternate compartments and spirals radially toward the weep

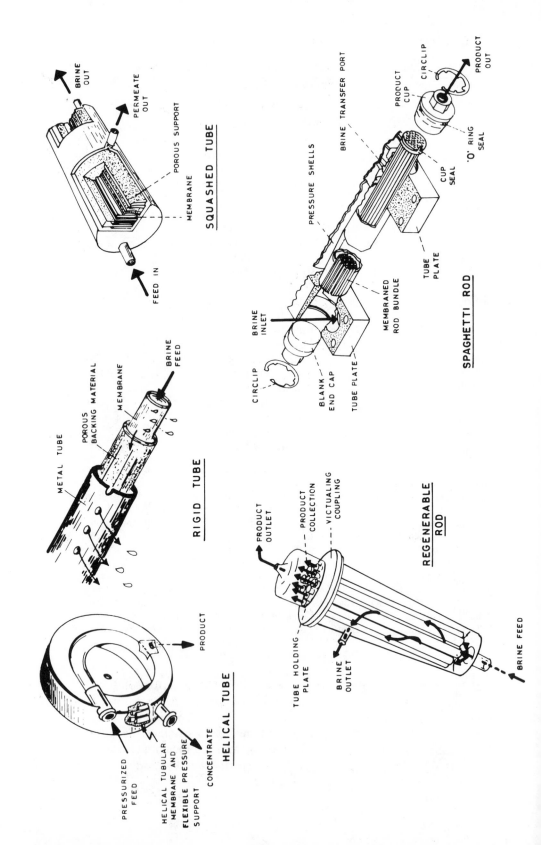

SQUASHED TUBE

BRINE OUT

PERMEATE OUT

POROUS SUPPORT

MEMBRANE

FEED IN

BRINE TRANSFER PORT

PRESSURE SHELLS

PRODUCT CUP

CIRCLIP

PRODUCT OUT

'O' RING SEAL

CUP SEAL

TUBE PLATE

MEMBRANED ROD BUNDLE

CIRCLIP

BRINE INLET

BLANK END CAP

TUBE PLATE

SPAGHETTI ROD

METAL TUBE

POROUS BACKING MATERIAL

MEMBRANE

BRINE FEED

RIGID TUBE

PRODUCT

HELICAL TUBE

PRESSURIZED FEED

HELICAL TUBULAR MEMBRANE AND FLEXIBLE PRESSURE SUPPORT

CONCENTRATE

PRODUCT OUTLET

PRODUCT COLLECTION

VICTUALING COUPLING

REGENERABLE ROD

TUBE HOLDING PLATE

BRINE OUTLET

BRINE FEED

holes in the central tube to be removed. Advantages of this design include fairly high water output per unit membrane area and vast amount of operating data.

The spiral-wrap design has probably been exposed to more volume of municipal effluent than any other design, aside perhaps from the rigid tubular design, and because of the small dimensions of the flow channel (see Table 10.8), it has a high probability of plugging. The design is one of the leading contenders for large-scale municipal treatment (Ajax International Corp., 1973). A similar design is also commercially available (see footnote *b* in Table 10.8).

3.3. Fiber Permeators (Figure 10.10)

Several million hollow fibers almost as fine as human hair (100–200 μm outside diameter) are bundled together in either a U-shape configuration (for brine flow on the outside) or in a straight configuration (for brine flow on the inside). The end of the fibers are epoxied into a tube sheet while making sure each fiber is not blocked. Thus, for the case in which the brine flows at high pressure on the outside of the hollow fibers, the product permeates radially inward through the unsupported fiber. The product then moves inside the hollow fiber bore to the product collection chamber. In the other case, where the design is similar to a typical heat exchanger, the brine flows into the bore of the hollow fibers at one end and, after moving along the inside of the fiber, flows out of the other end of the unit. The product continually permeates radially outward through the fiber walls.

The shell-side feed hollow-fiber design (i.e., brine on the outside) is very compact, low in cost, has a low water holdup, and because of the compressive strength of the small diameter fibers can withstand fairly high differential pressures (400 psi). It unfortunately plugs easily and is very hard to clean.

The inside-feed hollow-fiber design has the advantages of the shell-side feed plus the added advantage of well-controlled hydrodynamics of the feed, which improves the possibility of cleaning. This design is becoming very popular in biomedical and biochemical applications.

3.4. Plate-and-Frame Permeators (Figure 10.10)

The original ''plate-and-frame'' unit was similar in principal to the filter press (Aerojet General Corp., 1964, 1966). This design became defunct in the late 1960s because of many problems, the most important of which was the extreme difficulty and high expense of changing degraded membranes. Another unit using the flat-plate design is commercially available (Nielson,

---

Figure 10.9. Reverse osmosis membrane permeator designs. (Belfort, 1984.)

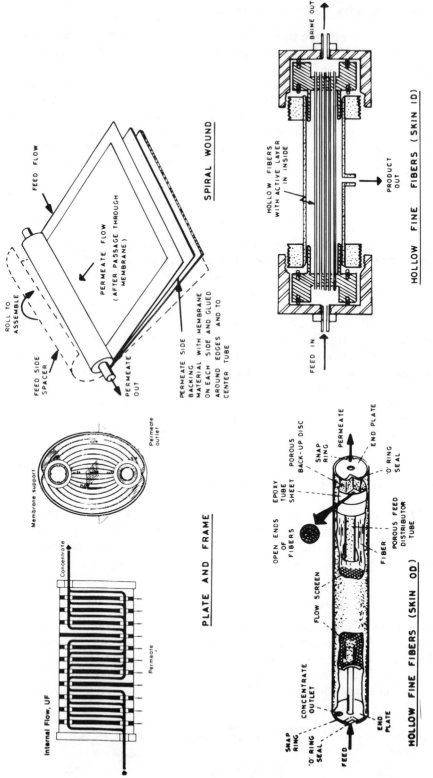

Figure 10.10. Reverse osmosis membrane permeator designs. (Belfort, 1984.)

TABLE 10.8. Comparison of Reverse Osmosis Membrane Permeators

Module Design	Packing Density (ft²/ft³)	Water Flux at 600 psi (gal/ft² day)	Salt Rejection	Water Output per Unit Volume (gal/ft³ day)	Flow Channel Size (in.)	Ease of Cleaning
Tubular						
1a Brine flow inside tube	30–50	10	Good	300–500	0.5–1.0	Very good
1b Brine flow outside tube[a]	140	10	Good	1400	0.0–0.125[a]	Good
Spiral wrap[b]	250	10	Good	2500	0.1	Fair
Fiber						
3a Brine flow inside fiber[c]	1000	5	Fair	5000	0.254	Fair
3b Brine flow outside fiber	5000–2500	1–3	Fair	5000–7500	0.002	Poor
Flat plate[d]	35	10	Good	350	0.01–0.02	Good
Dynamic membrane[e]	50	100	Poor	5000	~0.25	Good

Source: Belfort, 1984.

[a] Data for spaghetti permeator obtained from Grover et al. (1973). The flow channel dimension can vary from zero (tubes touching) to about 0.125 in.

[b] Two different spiral-wound designs are commercially available. In the one case (UOP, CA) the permeate spirals to the center manifold, whereas for the other design (Toray, Japan) the brine stream spirals to the center manifold. Their performances are essentially equivalent.

[c] Data for fiber with brine flow inside obtained from Strathman (1973). Maximum internal pressure for this unit is 28 atm (410 psi).

[d] Data for flat-plate design obtained from Nielsen (1972).

[e] Data for dynamic membrane design estimated from Thomas et al. (1973).

1972; Madsen et al., 1973). The original design was similar to a stack of phonograph records. After several years of intensive development, an improved design was developed and is now commercially available (Madsen, 1977). Alternate oval membranes and separating frames, which are also used for manifolding and sealing, are placed together and arranged for automatic internal staging in series. End plates are used to compress the stack as with a conventional filter press. This design does not need a cylindrical pressure vessel since each membrane is individually sealed by its neighboring separating frame. This is one of the main advantages over the original unwieldly plate and frame design. General advantages of this new design include the low brine holdup per unit membrane area and the ability to desalt highly viscous solutions because of the thin channel height (0.01–0.02 in.). Its disadvantages include susceptibility to channel plugging and difficulties in cleaning. With the new design, membrane replacement is extremely easy.

The other designs shown in Figures 10.9 and 10.10 and presented in Tables 10.7 and 10.8 are either not commercially available, such as the regenerable membrane designs, or have not made any impact on the market, such as the squashed membrane or the flat-plate spinning unit. The concept that a membrane can be regenerated in situ has been proven and reported in the literature (Belfort et al., 1973) but has not yet been commercialized.

Although most of the designs described above were intended for reverse osmosis applications, similar designs operating at lower applied pressures have been developed for ultrafiltration. The characteristics of three major UF-module designs are summarized in Table 10.9. In contrast to the reverse osmosis RO hollow fiber design with brine flow on the shell or outside of the fiber, the capillary or bore-side UF hollow-fiber design is easy to clean and can withstand the low operating pressure. Capillary, flat-plate, and spiral wound all look attractive for biotechnological applications for different reasons.

4. TRANSPORT EQUATIONS AND COEFFICIENTS

Since we are interested in the relative motion of various components through a membrane, it would be convenient to be able to describe this motion and thus be able to establish some basis for membrane performance. To attempt this, most researchers have invoked the theory of thermodynamics of irreversible processes (Prigogine, 1955). It is not our purpose here to develop this theory for membrane processes. Only the major results useful for discussion will be presented; for further details, we will provide various references. Before proceeding, however, two additional points should be made. The first is that the theory of thermodynamics of irreversible processes is a phenomenological description of the relative motion of various components within the membrane, which is itself considered to be "black box." This implies that the true microscopic mechanism of flow (and rejec-

TABLE 10.9. Comparison of Ultrafiltration Membrane Module Designs

Module Design	Packing[a] Density (m^2/m^3)	Water[b] Flux at 1 MPa ($l\ m^{-2}\ h^{-1}$)	Flow Channel Size (cm)	Volumetric Throughput per Unit Module Volume	Flow Regime	Shear Rate	Path Length (cm)	Ease of Cleaning
Flat plate	100–500	20–50	0.1–0.3	Medium	Laminar	Low	Short 30–50	Good
Tubular	100	20–50	1.27–2.54	Low	Turbulent	High	Long 3000	Very good
Spiral wrap[c] Capillary	1000–3000	20–50	0.1–0.6	High	Laminar	Low	Medium 50–200	Very good

[a] Taken from Manufacturer's Catalogs: DDS, Nakskov, Denmark and Millipore, Bedford, Mass., United States.

[b] 1.0 MPa ≡ 145 psi, pressure for an enzyme solution. Pure water fluxes can be as high as 5–8 times these volumes.

[c] Brine flows in the bore, and product is collected radially on the outside of the fiber.

tion) will not and cannot be explained by this theory. To the extent that this theory is combined with some "internal" membrane model, such as the solution–diffusion model in reverse osmosis, a mechanism can be inferred. The second point is that the theory of thermodynamics of irreversible process has been applied mainly to the pressure-driven as opposed to the other membrane processes.

Based on the theory of thermodynamics of irreversible processes, several approaches have been used to develop the basic transport equation, which relates the fluxes of solvents and solutes (V_i) to their respective driving forces (X_i) (Staverman, 1951; Spiegler and Kedem, 1966). These equations describe a coupling phenomenon that occurs between species when moving through the membrane. In general, for small deviations from equilibrium, we can write the following linear flux equations:

$$V_i = \sum_{j=1}^{m} L_{ij} X_j \qquad (i = 1, 2, \ldots, m) \tag{10.1}$$

Onsager (1931) has shown theoretically, and others have verified experimentally, that the following symmetry exists for the phenomenological coefficients:

$$L_{ij} = L_{ji} \qquad (i,j = 1, 2, \ldots, m) \tag{10.2}$$

Other restrictions on the coefficients are also operable and result from entropy considerations. They include

$$L_{ij} > 0 \qquad (i,j = 1, 2, \ldots, m) \tag{10.3}$$

and

$$L_{ii} L_j - L_{ij}^2 > 0 \qquad (i,j = 1, 2, \ldots, m) \tag{10.4}$$

The approach presented below uses the methods described by Haase (1969) to obtain the generalized equations for the isothermal heterogeneous (discontinuous) membrane system. Thereafter, two specific cases, such as for the reverse osmosis and UF processes, are examined. Thus, we will merely define the system and present the results obtained using the procedure described below.

Here, we consider two liquid subsystems separated from each other by a semipermeable membrane. Let the two homogeneous subsystems of our heterogeneous system be designated as phase' and phase" (see Table 10.10). According to this, we can attach a definite value for the pressure (P' and P''), for composition (molar concentrations C_k' or C_k''), and for the electrical potential (ϕ' or ϕ'') to each phase at constant temperature and at any arbitrary instant.

**TABLE 10.10. Heterogeneous (Discontinuous)
System Consisting of Two Homogeneous
Isotropic Subsystems (Phase′ and Phase″)**

Phase′	Phase″
Pressure P'	Pressure P''
Composition variable C'_k	Composition variable C''_k
Electrical potential ϕ	Electrical potential ϕ

Source: Belfort, 1984.

After performing mass, energy, and entropy balances across the membrane, an explicit expression of the dissipation function (of entropy) is derived. Then the fluxes (V_i) and forces (X_i) acting on the membrane system are chosen (by observation) so that all these quantities are independent and disappear at equilibrium.

For this system, without external forces, the following kinds of forces result at constant temperature (Haase, 1969):

$$X_i = \text{grad } \mu_i = v_i \text{grad } P + \left(\frac{\partial \mu_i}{\partial C_j}\right)_{T,P} \text{grad } C_j + z_i F \text{ grad } \phi \quad (10.5)$$

where v_i is the partial molar volume of species i, μ_i is the chemical potential of species i, z_i is the electrical charge on species i, F is Faraday's constant, and grad E refers to the gradient of a function E between phase″ and phase′.

Equation (10.5) together with some approximations can be used to develop the practical transport equations for various transport processes.

4.1. Reverse Osmosis and Ultrafiltration

Neglecting the last term (grad $\phi = 0$) in Eq. (10.5), and integrating across the thickness of the membrane, we get for the solvent (subscript 1) in a two-component system (and subscript 2 designates the solute)

$$\Delta \mu_1 = \int \left(\frac{\partial \mu_1}{\partial C_1}\right)_{P,T} dC_1 + \int v_1 \, dP = \int \left(\frac{\partial \mu_1}{\partial C_2}\right)_{P,T} dC_2 + \int v_1 \, dP \quad (10.6)$$

We know that when $\Delta \mu_1 = 0$, we are left with the effective osmotic pressure difference $\sigma \Delta \pi$ where σ, the Staverman reflective coefficient, represents the interaction between solute and solvent in the membrane. Thus, for constant v_1

$$v_1 \sigma \Delta \pi = -\int \left(\frac{\partial \mu_1}{\partial C_2}\right)_{P,T} dC_2 \quad (10.7)$$

and

$$\Delta\mu_1 = \nu_1 (\Delta P - \sigma\Delta\pi) \qquad (10.8)$$

For the solute,

$$\Delta\mu_2 = \int\left(\frac{\partial\mu_2}{\partial C_2}\right)_{P,T} dC_2 + \int\nu_2\, dP \qquad (10.9)$$

and for dilute solutions ($\mu_2' = \mu_2'^{\circ} + RT \ln C_2$) and constant ν_2, we get

$$\Delta\mu_2 = RT\Delta \ln C_2 + \nu_2\Delta P \qquad (10.10)$$

where the second term on the right-hand side of Eq. (10.10) is negligible with respect to the first for both pressure-driven processes (Belfort, 1977). Thus, we get

$$\Delta\mu_2 = RT\Delta \ln C_2 \approx \frac{RT}{C_2} \Delta C_2 \qquad (10.11)$$

and can incorporate $\Delta\mu_i$ (or X_i) into Eq. (10.1) by neglecting the cross efficients that are fairly small for the asymmetric cellulose acetate membrane (Bennion and Rhee, 1969) for solvent

$$V_1 = K_1(\Delta P - \sigma\Delta\pi) \qquad (10.12)$$

and for solute

$$V_2 = K_3\Delta C_2 = K_3 C_2' R \qquad (10.13)$$

where K_1 and K_3 are the water and solute permeability coefficients and are related to the phenomenological coefficients, ΔC_2 is the difference in solute concentration between bulk streams of product and feed, C' is the solute concentration of the bulk feed stream, and R is the coefficient solute rejection defined by

$$R = 1 - \frac{C_2''}{C_2'} \qquad (10.14)$$

To achieve greater accuracy, the concentrations at the membrane surface interfaces (C_{2m}' and C_{2m}'') should be used in Eqs. (10.13) and (10.14) instead of the bulk stream concentrations. Bulk stream values are, however, easier to measure and are usually used.

Thus, using equilibrium and irreversible thermodynamics and ignoring coupled flows, we have derived the flux equations. The resultant equations, assuming constant coefficients [Eq. (10.12) and Eq. (10.13)], are formulations of Fick's law of diffusion (Table 10.1). Thus, K_1 has been described in terms of a diffusion coefficient, water concentration, partial molar volume of water, absolute temperature, and effective membrane thickness. K_3 has been described in terms of a diffusion coefficient, distribution coefficient, and effective membrane thickness (Merten, 1966).

For ultrafiltration and microfiltration, where the osmotic pressure is usually negligible relative to the applied pressure ($\sigma\Delta\pi < \Delta P$), we get for solvent

$$V_1 = K_1\Delta P \tag{10.15}$$

and for solute

$$V_2 = K_3\Delta C_2 = K_3\, C_2'R \tag{10.16}$$

5. CONCENTRATION POLARIZATION

As mentioned in the Introduction, concentration polarization severely limits transmembrane throughput (10-fold or greater reduction relative to pure water throughput) under normal operating conditions (Michaels, 1981); see Table 10.11. Ingenious designs have been developed to house membranes in such a way that maximum performance could be attained by minimizing concentration polarization (Table 10.12). Thus, designs with rotating disk membranes, vibrating spaghettilike membrane-support structures, turbulence parameters or mixers inserted above the membrane surface, and ultrathin channel devices have been developed. Fluid management techniques have been used to diminish its depreciating effects (Table 10.13). Operating innovations such as pulsating and periodic reverse flow have also been used.

Extremely important factors, usually not well undrstood and therefore ignored, are the physical and chemical surface characteristics of the membrane. See Table 10.14.

To understand concentration polarization, the fundamental equations have to be formulated and solved. This has been done under various assumptions for unstirred batch, laminar flow, and turbulent flow systems. Typical assumptions for hyperfiltration are that the membranes are ideal (i.e., $R = 1$), feed concentration is low (i.e., $\Delta\pi \ll \Delta P$), flow is fully developed, viscosity is independent of concentration, diffusivity is dependent on concentration, and so on. Confirmation of most of these models has been through measurements, but very few models have been compared with measurements. Examples of direct measurements include Hendricks and William's (1970) method, which has two disadvantages. The probe may disturb the observation, and the probe head was at least the size of and larger than the total thickness of the steady-state polarization layer. Mahlab

TABLE 10.11. Limitations of Ultrafiltration

Major limitation: *Solute polarization*	1. Build up of retained species (macromolecules, sub-micron colloids) at the membrane–solution interface.
Possible effects *of polarization*	1. Increased osmotic pressure reduces driving force and flux. 2. Solute–surface interactions[a] that could plug surface porosity and reduce flux. 3. Solute–solute interactions[b] resulting in aggregation and possible destabilization and cake build up reduces flux. 4. Biological film growth may increase or reduce flux.
Important factors	1. Nature of the various surfaces and interactions. 2. Conditions of the solution (pH, ionic strength, electrolyte composition, temperature, pressure, and solute concentrations). 3. Solute characteristics (compressibility, charge).
Results	1. Reduced flux. 2. Effect retention relationships.
Approaches to reduce *polarization*	1. Optimize design of module. 2. Manage hydrodynamic flow. 3. Pretreat membrane to reduce solute–membrane interactions. 4. Pretreatment of feed. 5. Periodic membrane cleaning.

[a] Resulting in adsorption, ion exchange.
[b] Resulting in precipitation and/or gel formation.

and co-workers (1980) have used the nondisturbance method of laser interferometry with ray tracing to account for beam deflection. Not only was this technique used for transient analysis but also for studying membrane compaction and rejection of various salts and urea (Mahlab et al., 1981).

For comprehensive review of the current state of the art, see Matthiasson and Sivik's publication (1980). Figure 10.11 presents a comprehensive scheme for reducing concentration polarization (Bruin et al., 1980).

Since concentration polarization represents a concentrating process (bulk to wall), some have suggested that perhaps it could be used positively. Lee and Lightfoot (1974) have suggested skimming the concentrated layer, whereas others have concentrated enzymes in the polarization layer and have used the device as a reactor.

6. FOULING

One of the major problems encountered in pressure-driven membrane separation systems such as reverse osmosis and ultrafiltration is the build up of

TABLE 10.12. Design Suggestions to Reduce Concentration Polarization (cp)

Design for	Effect	Result
1. Low flux high surface area—hollow fibers (outside flow)	Reduced polarization due to convection, but because of low axial flow cp is increased.	Excellent steady performance for "clean" solutions
2. Thin channel, short path lengths—flat membrane and capillary modules	Reduces boundary layer thickness increasing back diffusion. Operate in laminar-flow conditions and entrance effects important, low shear condition.	Lower pumping energy, higher fluxes
3. Minimum energy losses in axial feed flow for operation in turbulent region—tubular modules	Turbulent eddies increase back diffusion of solute, high shear condition.	Higher fluxes and pumping energy
4. Spinning membrane to reduce cp. Flat membrane disks or cylindrical membranes (outside feed) are rotated	Reduces the solute build up due to density differences between the polarized layer and bulk solution. Taylor vorticies induce secondary flows.	Higher fluxes but increased energy needed for rotation
5. Compromise between low-flux/high-surface area (1) and high-flux/low-surface area—use designs such as the capillary (inside flow of feed) or spiral-wound modules (inside or outside feed flow)	Reduced polarization due to screens or mixing parameters with medium high packing density.	Excellent compromise between (1) and (3) above

a flux-reducing fouling layer on the membrane surface. Adequate modeling of this phenomenon is an important step in understanding the physics of the problem and in developing improved operating techniques and module designs.

In practical systems, maintenance of an economically acceptable flux is usually obtained by costly pretreatment and cleaning procedures. By choosing the appropriate membrane type, module design, and fluid flow regime, optimum performance can be obtained at minimum cost. Usually, the choice of membrane will depend on the flux-retention characteristics desired, the environmental state of the feed (temperature and pressure), and the characteristics of the potential foulants in the feed solution. For example, hydrophobic-type membrane surfaces such as polysulfone show high protein ad-

TABLE 10.13. Hydrodynamic Methods to Reduce Concentration Polarization (cp)

Method	Effect	Result
Reduce convective flux	Reduces rate of cp build up	Lower flux (and volumetric flow rate)
Use tangential feed flow rather than impact (direct-on flow)	Increase the back diffusion of polarized species, reducing the build up of solute (decreased osmotic pressure and gel layer resistance)	Higher flux
Introduce frequent feed flow reversal	Reduces solute build up at the entrance region by changing the entrance region position.	Higher flux
Introduce mixing promotors into the feed channel	Increases boundary layer mixing thereby increasing the back diffusion rate. Both fixed and moving (fluidized) inserts have been used.	High fluxes and higher pressure drop (pumping energy)
Use frequent pressure pulses of the feed flow	Destabilizes the cp build up and causes increased back diffusion.	Higher flux
Addition of dynamic particles to feed stream	Because of the well-known "tubular-pinch" effect, added particles will, if chosen correctly, disturb the cp layer increasing back-transport.	Higher flux

sorption, whereas hydrophilic-type membrane surfaces such as cellulose acetate or polyamide exhibit much lower protein adsorption. Which module design to choose will also depend on the feed solute characteristics, volumetric flow rate desired, and, as discussed earlier, the appropriate channel configuration. For example, excellent hydrodynamics control is possible with tubular, capillary, or sheet flow systems. On the other hand, the presence of large suspended particles may plug or block the narrow flow channel in capillary or sheet flow.

The main purpose in modeling the feed-flow conditions and the fouling process in membrane ducts is for predicting the effect on permeation flux due to changes in axial flow rate, feed concentration, and applied pressure differential. Most fouling models have been developed for the retention of macromolecules, although several recent attempts have been made with colloidal solutions.

TABLE 10.14. Chemical and Physical Membrane Methods to Reduce Concentration Polarization (cp)

Method	Effect	Result
Chemical		
Change nature of membrane surface to be polar, nonpolar, and hydrophillic or hydrophobic	Changes the solute–membrane (sm) surface interaction causing either attraction or repulsion of polarized solute. Using electrostatic attraction, solutes could later be removed by competitive charge species (pH, other ions).	1. By increasing sm interactions, solute could be removed from the stream for later recovery. 2. By decreasing sm interactions, solute could be repelled from the surface reducing flux decline.
Physical		
Physically cast the membrane with protrusions	Increases mixing in the mass transfer boundary layer.	Increases flux.

6.1. General Equations

The coupled nonlinear equations of change are given by

$$\frac{D\rho}{Dt} = -\rho \, (\bar{\nabla} \cdot \bar{v}) \text{ (fluid continuity)} \tag{10.17}$$

$$\rho \, \frac{D\bar{v}}{Dt} = -\bar{\nabla}p - \bar{\nabla} \cdot \bar{\bar{\tau}} + \rho\bar{g} \text{ (fluid motion)} \tag{10.18}$$

$$\frac{DC}{Dt} = D\bar{\nabla}^2 C - \bar{v} \cdot \bar{\nabla}C \text{ (solute continuity)} \tag{10.19}$$

and

$$q = f(c) \text{ (adsorption isotherm)} \tag{10.20}$$

6.2. Fluid Flow in Porous Ducts

Exact solutions for the low Reynold's number flow of an incompressible Newtonian fluid in tubes in slits with permeation through the duct walls is given by Kozinski et al. (1970).

The tube has inside dimension of radius $r = R$ and axial flow along the z direction, whereas the slit has half-height b in the y direction and axial flow along the x direction. The width w is in the z direction. The complicated equations in Kozinski et al. (1970) can be simpled for the special case of

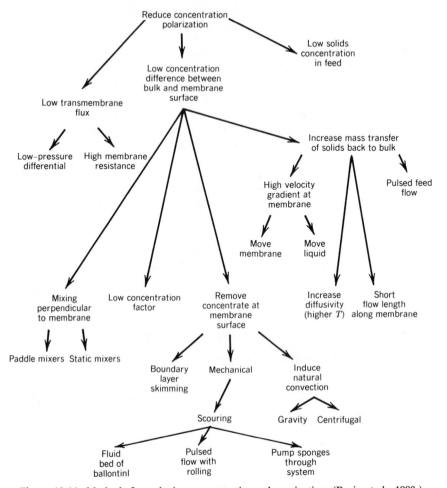

Figure 10.11. Methods for reducing concentration polymerization. (Bruin et al., 1980.)

limiting seepage rate, that is, $\alpha R \ll 1$ and $\alpha b \ll 1$, respectively. Thus, for the tube

$$v_z = v_z(o,z)\left[1 - \left(\frac{r}{R}\right)^2\right] \tag{10.21}$$

$$V_z(o,z) = \frac{2Q_0}{\pi R^2} - \frac{4V_o}{\alpha R}(1 - e^{-\alpha z}) \tag{10.22}$$

$$V_r(r,z) = V_r(R,z)\left[2\left(\frac{r}{R}\right) - \left(\frac{r}{R}\right)^3\right] \tag{10.23}$$

$$V_r(R,z) = V_o e^{-\alpha z} \tag{10.24}$$

and for the slit

$$V_x(y,x) = V_x(o,x) \left[1 - \left(\frac{y}{b} \right)^2 \right] \tag{10.25}$$

$$V_x(o,x) = \frac{3}{4} \frac{Q_0}{wb} - \frac{3V_o}{2\alpha b} (1 - e^{-\alpha z}) \tag{10.26}$$

$$V_y(y,x) = \frac{v_y(b,x)}{2} \left[3 \left(\frac{y}{b} \right) - \left(\frac{y}{b} \right)^3 \right] \tag{10.27}$$

$$V_y(b,x) = V_o e^{-\alpha z} \tag{10.28}$$

where for a kidney tubule $\alpha R \sim 10^{-3}$ and the above equations hold.

6.3. Unstirred Batch Cell (No Flow)

Mahlab et al. (1980) have derived theoretically and measured interferometrically the concentration profile for dilute dissolved solutes adjacent to a hyperfiltration (reverse omosis) membrane. For the case of rejection $R = 1 - C_p/C_o = 1$, the polarization parameter $\psi = C/C_o - 1$ is given by

$$\psi_{k=1}(o,\tau) = 1 + \tau + \sqrt{\tau/\pi} \, e^{-\tau/4} - (1 + \tau/2) \, \text{erfc} \, (\sqrt{\tau/2}) \tag{10.29}$$

As $\tau = V_{w_o}^2 t/R \to \infty$ for $R < 1$, the steady-state solution obtained was

$$\psi_{r<1}(\xi,\infty) = \frac{R}{1 - R} e^{-\xi} \tag{10.30}$$

where the dimensionless axial parameters $\xi = V_{w_o} x/D$ and V_{w_o} equal initial wall flux at $x = 0$.

Trettin and Doshi (1980) have performed a similar analysis using similarity transformation for ultrafiltration of macromolecular solutions. Several workers have attempted to explain the experimental observation of pressure-independent flux by suggesting gel limitations (Michaels, 1968; Blatt et al., 1970). Others have also proposed that osmotic effects could be used to explain this phenomenon. Using

$$|V_w| = \frac{\Delta P - \sigma \Delta \Pi}{\mu (R_m + R_g)} \tag{10.31}$$

Blatt et al. (1970) suggested that under gel limitation where $\Delta P \gg \sigma \Delta \Pi$ and $R_g \gg R_m$, Eq. (10.31) reduces to

$$|V_w| \simeq \frac{\Delta P}{\mu R_g} \tag{10.32}$$

Thus, any increase in ΔP would result in a concomitant increase in gel layer thickness (and hence R_g), resulting in a relatively constant $|V_w|$. If on the other hand, before gel formation ($R_g = 0$) the polarization of solute at the membrane–solution interface results in an increase in $\Delta\Pi$ due to an increase in ΔP, then the flux

$$|V_w| = \frac{\Delta P - \Delta\Pi}{\mu R_m} \qquad (10.33)$$

could appear pressure independent because of constant $\Delta P - \Delta\Pi$.

Trettin and Doshi (1981) have analyzed these two limiting problems for unstirred batch cell and for a thin channel cross-flow system. They obtained

$$\frac{\Delta V_{exp}}{T^{1/2}} = \left(\frac{\Delta V}{T^{1/2}}\right)_{lim} + \Delta V_{corr}\left(\frac{1}{T^{1/2}}\right) \qquad (10.34)$$

for unstirred batch cell and

$$\frac{|\dot{Q}_p|_{exp}}{L^{2/3}} = \left(\frac{|\dot{Q}_p|}{L^{2/3}}\right)_{lim} + |\dot{Q}_p|_{corr}\left(\frac{1}{L^{2/3}}\right) \qquad (10.35)$$

for cross-flow systems. ΔV is the accumulated permeation volume in time T, and $\dot{Q}_p$ is the average volumetric permeation rate. The corrected terms account for the initial period of filtration when $C_w < C_{wa}$ (osmotic limited maximum) or C_g (gel limited). From a plot of $(\Delta V_{exp}/T_{1/2})$ versus $(T^{-1/2})$ or $(|\dot{Q}_p|_{exp}/L^{2/3})$ versus $(L^{-2/3})$, the limiting or infinite time extrapolation case is obtained. If $[\Delta V/T^{1/2}]_{lim}$ or $[|\dot{Q}_p|/L^{2/3}]_{lim}$ are pressure dependent, clearly the system is osmotically limited; whereas if these groups are pressure independent, the system is gel limited. In Figure 10.12, obtained from Trettin and Doshi (1981), these two effects are observed for the ultrafiltration of bovine serum albumin in an unstirred batch UF cell.

For high-value products, batch processing and separation are sometimes used. However, mixing is usually conducted on the feed side to reduce the concentration polarization effects.

6.4. Cross Flow (Continuous)

6.4.1. Gel-Polarization Model

Michaels (1968) and later Blatt et al. (1970) presented and analyzed the gel-polarization (GP) model. It involves a solute mass balance over a thin volume slice at the membrane–solution interface for a batch or cross-flow continuous cell at position z, as shown in Figure 10.13.

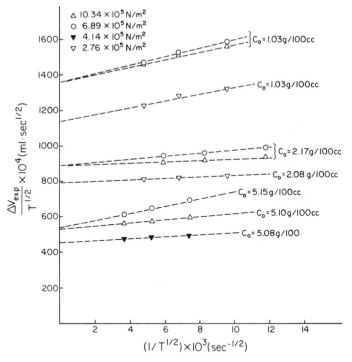

Figure 10.12. Unstirred batch cell UF of 0.15 M NaCl BSA solution ($T = 21–24°C$, pH 7.4) at various solute concentrations and applied pressures. (Trettin and Doshi 1981.)

$$\text{Mass flux to interface} = \text{Mass flux away from interface} \quad (10.36)$$

$$V_w C = D \frac{dc}{dr} + V_w C_p$$

Separating variables and integrating across the mass boundary layer thickness δ, gives

$$V_w = \frac{D}{\delta} \ln \frac{C_w - C_p}{C_b - C_p} \quad (10.37)$$

If $R = 1 - C_p/C_w$, then

$$\frac{C_w}{C_b} = \frac{\exp(V_w \delta / D)}{R + (1 - R) \exp(V_w \delta / D)} \quad (10.38)$$

If $R = 1$, that is, $C_p = 0$, then

$$V_w = \left(\frac{D}{\delta}\right) \ln \left(\frac{C_w}{C_b}\right) \quad (10.39)$$

I. BATCH (impact or dead ended)

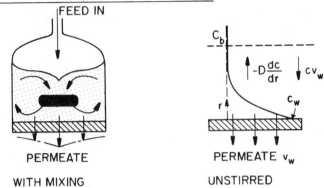

PERMEATE   PERMEATE v_w

WITH MIXING   UNSTIRRED

2. CROSS-FLOW CONTINUOUS

MASS TRANSFER IN POROUS DUCTS WITH SUCTION

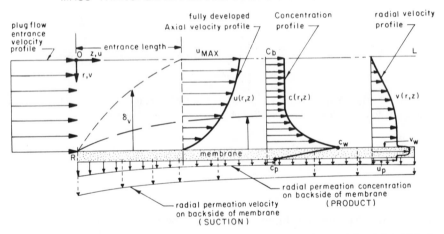

Figure 10.13. Two different flow methods above a membrane surface. (*a*) Batch in which the bulk fluid flows toward the membrane and (*b*) cross flow in which the bulk fluid flows tangentically across the membrane. Characteristic velocity and mass profiles are shown.

or

$$Pe'_w = \frac{V_w \delta}{D} = \ln \frac{C_w}{C_b} \tag{10.40}$$

The film mass transfer coefficient $k = D/\delta$ is usually obtained from mass-transfer correlations for laminar and turbulent flow in *nonporous* ducts. For flow in tubes the following equations are applicable.

Laminar

$$Sh = 1.295 \left(\frac{R}{L} ReSc \right)^{1/3} \tag{10.41}$$

where

$$Sh = \frac{k(2R)}{D}; \qquad Re = \frac{2R\bar{U}}{\nu}; \qquad Sc = \frac{\nu}{D}$$

or

$$k = \frac{D}{\delta} = 0.816 \left(\frac{\dot{\gamma}}{L} D^2 \right)^{1/3} \tag{10.42}$$

$$\dot{\gamma} = \frac{\bar{U}}{R} = \frac{U_{max}}{2R}$$

where L is the path length, Re is Reynolds number, defined above

Turbulent

$Sh = 0.023 \, Re^{0.83} \, Sc^{0.33}$ (theoretical) $\tag{10.43}$
$Sh = 0.023 \, Re^{0.875} Sc^{0.25}$ [experimental (Nakao and Kimura, 1981)] $\tag{10.44}$

with

$$\frac{f}{2} \equiv 0.023 \, Re^{-0.2} \tag{10.45}$$

and

$$Sh = \frac{f}{2} Re^{1.03} Sc^{0.33} \tag{10.46}$$

Three significant and useful results are obtained from the GP model [Eq. (10.39)] together with the k correlations [Eqs. (10.42)] and (10.43)]: The flux is proportional to the ln C_b^{-1}; the flux is invarient with applied pressure; and the flux is proportional to the average axial velocity $\bar{U}$ to the 0.33 power (laminar) and to the 0.83 power (turbulent). These relationships are shown schematically in Figure 10.14. Ample experimental evidence is available for dissolved macromolecules supporting the GP model and the relationships shown in Figure 10.14. Clearly, for pressure independence the model only applies for solute concentrations $C > 0$ and at $P > P_{max}$. Below P_{max}, as $P \to$ 0 Darcy-type behavior applies between V_w and P as shown in Figure 10.14a. The convergence of linear plots to ln C_w in Figure 10.14b is a test of the

GP MODEL - HYDRODYNAMIC DIAGNOSTICS

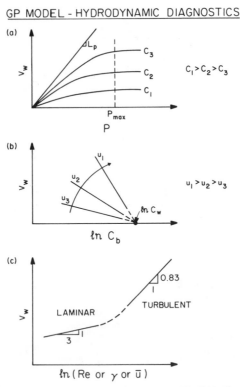

Figure 10.14. Typical flux (v_w) versus (a) applied pressure (P), (b) bulk solute concentration (ln C_b), and (c) axial cross flow (Reynolds number N_{Re}, shear γ, or average axial velocity, $\bar{u}$).

model, whereas the slope of flux V_w versus flow rate in Figure 10.14c confirms the usefulness of the friction factor correlations for laminar and turbulent flow.

Experimental evidence supporting the trends shown in Figure 10.14 will be presented below for macromolecular solutions. Porter (1972) has presented data for bovine serum albumin solutions shown in Figure 10.15a,b,c. The similarity between Figures 10.14a,b,c, and Figures 10.15a,b,c is clear. This model has thus been very useful as a diagnostic tool for understanding and predicting the fouling behavior of macromolecular solutions.

For ultrafiltration of macromolecules, Probstein and co-workers (1979) have proposed the following improvements to the GP model:

For laminar flow (Re $\leq$ 700):

$$V_{w_{\lim}} = 1.31 \left(\frac{U D_g^2}{hL} \right)^{1/3} \ln \frac{C_g}{C_b} \qquad (10.47a)$$

or

$$V_{w_{\lim}} \left(\frac{h^2}{Q}\right)^{1/3} \approx \ln C_g - \ln C_b \qquad (10.47b)$$

where Q is the flow rate and D_g is gel diffusivity.

For turbulent flow (Re $\geq$ 1100):

$$V_{w_{\lim}} = 0.011 \frac{D_g}{h} \, Sc^{0.25} \, Re^{0.9} \ln \frac{C_g}{C_b} \qquad (10.48a)$$

or

$$V_{w_i_m} Re^{-0.9} \sim \ln C_g - \ln C_b \qquad (10.48b)$$

where $Sc^{1/4}$ is weak and hence neglected. Equations (10.47b) and (10.48b) are used to obtain C_g. Note that the exponent of the Reynold's number is close to that measured by Nakao and Kimura (1981) [see Eq. (10.44)].

Using a tubular system, Nakao and Kimura (1981) have conducted extensive experimental studies on macromolecular ultrafiltration. Their experimental results were expressed in terms of Eq. 10.44, confirming the turbulent model (Eq. 10.43). Their data, however, did not follow the GP model with respect to Figure 10.14b. They found that R_g (Eq. 10.31) was a function of C_w in equating Eqs. 10.31 with 10.37.

For cross-flow membrane plasmapheresis, Zydney and Colten (1982) have presented a variant of the GP model using Eckstein and co-worker's (1972) correlation for "shear-enhanced" diffusion coefficient. By substituting for D from ($C_b > 0.20$)

$$D = 0.025 \, a^2 \dot{\gamma} \qquad (10.49)$$

into the mass transfer coefficient $k(z) = D/\delta$ [Eq. (10.42)] and then substituting this into Eq. (10.39), one obtains the wall flux as a function of axial distance, z,

$$V_w = 0.047 \left(\frac{a^4}{z}\right)^{1/3} \dot{\gamma}_w \ln \frac{C_w}{C_b} \, . \qquad (10.50)$$

Integrating along the length L, one obtains

$$\bar{V}_w = \frac{1}{L} \int_0^L V_w(z) \, dz = 0.070 \left(\frac{a^4}{L}\right)^{1/3} \dot{\gamma}_w \ln \frac{C_w}{C_b} \qquad (10.51)$$

where $\bar{V}_w$ is the length-averaged flux. A value of $C_w = 0.95$ (v/v) was used for plasma filtration (Zydney and Colton, 1982). Results comparing the

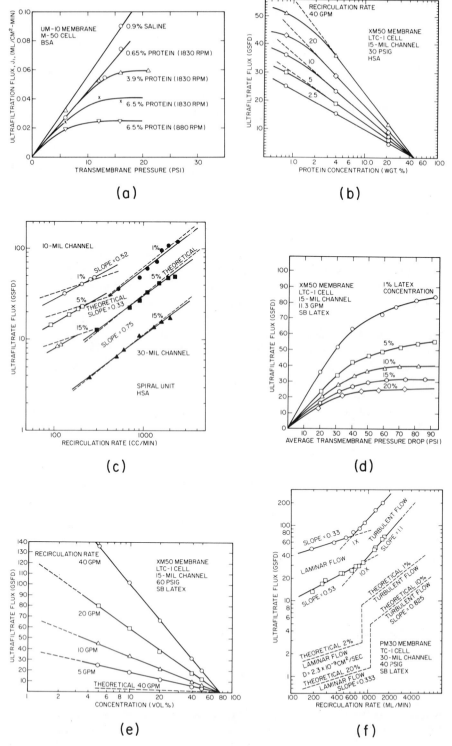

(a)

(b)

(c)

(d)

(e)

(f)

phenomenological model of Eq. (10.51) with measured plasma filtrate flux
are shown in Figure 10.16. The fit looks good; however individual sets of
data do not follow the theoretical model as well.

Blatt et al. (1970) and later Porter (1972), Madsen (1977), and others
recognized that the GP model, whereas excellent for predicting the behavior
of macromolecular solutions, under predicted the flux behavior for particulate or colloidal suspensions. See Figure 10.15d,e,f for colloidal suspensions. This latter phenomenon has been termed "the flux paradox for colloidal suspensions" by Green and Belfort (1980) and is clearly seen in Figure
10.15f. Although Porter (1972), Madsen (1977), Thomas and Gallaher (1973),
and Gutman (1977) have all hypothesized that the observed increased flux
could result from augmented back migration of particles away from the
fouling film toward the bulk solution resulting from the so-called "tubular-
pinch effect" (Segre and Silberberg, 1981; 1962), none has provided a satisfactory quantitative model incorporating particle lift forces to account for
this effect.

6.4.2. Gel-Polarization Lateral-Migration Model

As a consequence of some complicated mathematical arguments, Altena and
Belfort (1984), Belfort and Altena (1983), and Altena et al. (1983) have shown
that when

$$V_w \sim \frac{\rho a^3 \, \bar{U}^2}{\mu d^2} \tag{10.52}$$

the lift velocity V_L resulting from inertial effects is effective in counteracting
the V_w drag of neutrally bouyant colloids toward a membrane. Belfort et al.
(1982) derived the following flux equation (in the limit as $\delta \ll d$) analogous
to the GP model [Eq. (10.39)]:

$$V_w(\delta \ll d) = \left[1.295 \left(\frac{D^2}{dL} \right)^{1/3} \ln \frac{C_w}{C_b} \right] \bar{U}^{1/3} + \alpha \kappa^2 \left(\frac{a}{v} \right) \left(1 - \frac{r^*}{d} \right) \bar{U}^2 \dots$$

$$\underbrace{\hspace{4.5cm}}_{\text{Diffusive term}} \qquad \underbrace{\hspace{5cm}}_{\text{Lateral migration term}}$$

$$\tag{10.53}$$

where d = tube radius, $\kappa = a/d$, a = particle radius, r^* = equilibrium
position toward which the particles migrate, v = kinematic viscosity, and α
= empirical constant approximately equal to 5 (Ishi and Hasimoto, 1980).
Notice that the first term on the right-hand side of Eq. (10.53) is identical to

---

Figure 10.15. Confirmation of the GP model for macromolecules such as serum alburim (a–c)
and the flux paradox for suspended particles such as styrene butadiene laticies (d–f). (Porter,
1972.)

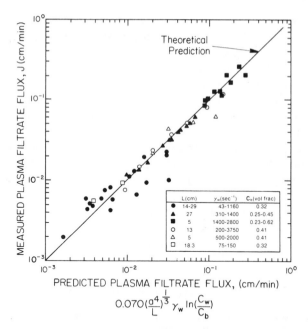

Figure 10.16. Comparison between predicted and measured plasma filtrate flux. Data for tube (●, □) and flat plate (△, ▲, ■, ○) devices. Wall shear rate evaluated at average value, bulk concentration at inlet value (taken from Zydney and Colton, 1982.)

the GP model, whereas the second term incorporates lateral migration. Thus, for macromolecular solutions ($a \to 0$) where lateral migration is probably small, the first-term dominates in Eq. (10.53). For purely colloidal suspensions, however, the reverse is true (D is very small) and the lateral migration term probably dominates. For mixed solutions, each term will probably contribute to the flux. Thus, a plot of flux versus average axial velocity could have a power slope from ⅓ to 2.

Experimental evidence supporting this model is shown in Figures 10.17a,b,c, where the power slopes of the flux versus axial velocity are greater than ⅓ as predicted by the gp model. In Figure 10.17c the power slope of plasma only is about 0.33, whereas that for whole blood (containing red blood cells) is about 0.6. Using wastewater and tap water, Thomas and Gallaher (1973) and Reed and Belfort (1982) obtained a power slope of 2 (Fig. 10.17b).

Additional results from Porter (1972) with styrene–butadiene latex particles show power slopes of about 0.8 for laminar flow (Fig. 10.17a).

In summary, the new gel-polarization lateral-migration (GPLM) model is able to predict the flux–axial velocity relationship for macromolecular solution, colloidal suspensions, and mixtures thereof. The wall concentrations C_w, solute diffusivity in the bulk (D) or in the gel (D_g), plus the equilibrium position r^* are all needed for an a priori prediction of permeation flux.

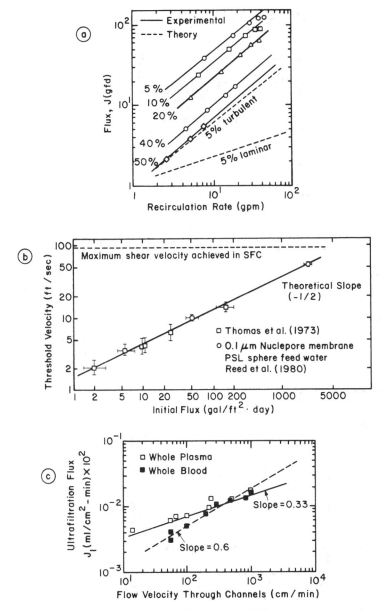

Figure 10.17. Experimental results showing flux versus axial flow rate across membranes with and without suspended particles. (*a*) Slopes higher than expected for laminar flow of 0.1 μm laticies (Porter, 1972); (*b*) slope approximately equal to 2 (inverse of ½) for ultrafiltration of wastewater (Thomas and Gallaher, 1973) and microfiltrating polystyrene spheres (Belfort et al., 1982); and (*c*) change of slope from 0.33 to 0.6 on adding erythrocytes to whole plasma (Porter, 1972).

6.4.3. Resistance Model

In another approach in which the $r^* \to d$ such that permeation drag is greater than lift drag, that is, $V_w > \rho a^3 \bar{U}^2 / \mu d^2$ [from Eq. (10.52)], standard filtration theory has been adapted to membrane processes by Belfort and co-workers (Green and Belfort, 1980; Belfort and Marx, 1979). This method provides an estimate of the cake and membrane resistances. The pressure-driven flux through the membrane and fouling layers in series is given by (neglecting osmotic effects):

$$V_w = \frac{\Delta P}{\mu(R_c + R_m)} \qquad (10.54)$$

or

$$V_w^{-1} = \frac{\mu R_c}{\Delta P} + \frac{\mu R_m}{\Delta P} . \qquad (10.55)$$

From the standard cake filtration equations the cake resistance is given by

$$R_c = \left(\frac{\alpha^\circ w}{2A}\right) V = \alpha'' V . \qquad (10.56)$$

The accumulated permeation volume

$$V = \int_0^t A V_w' \, dt \qquad (10.57)$$

depends on V_w' which is the <u>net</u> permeation drag, that is, when inertial lift drag is of the order of permeation drag and Eq. (10.40) holds, $V_w' = V_w - V_L$ with V_L given by

$$V_L = -\alpha \kappa^2 \left(\frac{\bar{U}^2 a}{\nu}\right) \left(\frac{r}{d}\right) \left[\left(\frac{r^*}{d}\right) - \left(\frac{r}{d}\right)\right] . \qquad (10.58)$$

Using the Carman-Kozeny equation for R_c,

$$R_c = \frac{180 (1 - \epsilon)^2 \delta_c}{(2a)^2 \epsilon^3} \qquad (10.59)$$

for a close-packed spherical particulate cake with porosity ϵ. A mass balance over the cake obtains

$$\delta_c = \frac{c_f}{\rho_p(1 - \epsilon)} \int_0^t (V_w - V_L) \, dt \qquad (10.60)$$

A stepwise iterative procedure to calculate the permeation flux V_w as a function of time can be formulated with a detailed knowledge of the flow conditions and suspension properties. In step 1, the membrane resistance R_m is obtained from the initial flux, driving pressure, and viscosity, with $R_c = 0$ in Eq. (10.54). Knowing the channel (R_T) and colloid dimensions (r_p), the average flux velocity $\bar{U}$, V_L is calculated using Ishi and Hasimoto's (1980) empirical relation or Ho and Leal's (1965) theoretical model. In step 3, R_c is calculated with V_w and V_L substituted into Eq. (10.60) for δ_c and then into Eq. (10.59), allowing a new V_w to be calculated from Eq. (10.54). These three steps are now recalculated with the revised V_w at $t + \Delta t$ and a new channel radius of $R_T - \delta_c$. Three examples using this procedure are shown in Figure 10.18. For the same feed concentration c_f, the flux decline is arrested earlier for the smaller duct radius than the larger duct radius. This explains why the narrow thin-slit membrane module has proved to be so attractive for highly turbid waters. For the same duct radius, however, the feed with the lower feed concentration has a lower initial flux decline reaching the steady flux later than the case with the higher feed concentration.

6.4.4. Test of the Resistance Filtration Model

Madsen (1977) claims that application of the lift velocity expressions to colloidal ultrafiltration results in predicted steady-state fluxes that are considerably lower than those observed. Madsen, however, applied the lift velocity expressions in an unsatisfactory manner. The following is a brief

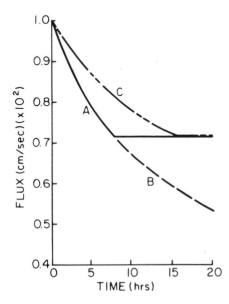

Figure 10.18. Flux-decline plots using iterative procedure described in text for (a) $R_T = 0.1$ cm, $C_f = 250$ ppm; (b) $R_T = 1.0$ cm, $C_f = 250$ ppm; and (c) $R_T = 0.1$ cm, $C_f = 125$ ppm. Other conditions were constant and $\Delta p = 100$ psi, $\mu = 0.01$ poise, Re $= 100$, $r_p = 0.5$ μm, $\epsilon = 0.3$, and $J_{initial} = 1 \times 10^{-2}$ cm sec^{-1}. (Green and Belfort, 1980.)

outline of how the lift expressions should, in fact, be applied to ultrafiltration modeling.

When fouling occurs, that is, $V_w > V_L$, the growth of a cake on the membrane serves to reduce the effective tube radius (or slit height) of the module. This reduction in the effective tube radius has a large effect on the lift velocity, as can be seen from Eq. (10.58) where lift velocity is inversely proportional to tube radius to the third power and proportional to the square of the average axial velocity $\bar{U}^2$, viz

$$V_L \propto \frac{\bar{U}^2}{R_T^{\,3}} \tag{10.61}$$

For a cake thickness δ_c replacing tube radius R_T by the actual slit half-height $(B - \delta_c)$ and realizing that for a constant displacement pump

$$\bar{U} \propto \frac{Q}{[2(B - \delta_c)w]} \tag{10.62}$$

for a slit, the following results:

$$V_L \propto \frac{Q^2}{[4w^2(B - \delta_c)^5]} \tag{10.63}$$

Thus, slight decreases in the effective slit half-height yield very large increases in the lift velocity. The hydraulic resistance to flux can be obtained from Eqs. (10.54) and (10.56):

$$V_w = \frac{\Delta P}{\left[\mu\left(\dfrac{\alpha w^{\circ} V}{2A}\right) + R_m\right]} \tag{10.64}$$

where $w^{\circ}V/A$ is proportional to cake build-up thickness δ_c ($k^* = $ proportionality constant)

$$V_w = \frac{\Delta P}{[\mu(k^* \delta_c + R_m]} \tag{10.65}$$

Thus, the sensitivity of V_L is far greater than V_w to change in δc. Steady-state flux is achieved when V_L and V_w become equal in magnitude. Steady-state flux can only be predicted by evaluating the lift velocity at the steady-state effective tube radius, R_T, not at the original tube radius, R_{T_0}, as was done by Madsen.

Thus, slight differences in effective tube radius, neglected by Madsen (1977), provide the margin between correct and grossly incorrect predictions of steady-state flux.

Using the above equations and the iterative procedure described previously, the dynamic growth of the fouling layer is followed with time through the flux V_w and lift velocity V_L in Figure 10.19 for two types of prime movers. The data were taken from Porter's paper (1972). The data concern the ultrafiltration of styrene–butadiene polymer latex in a channel of width 1.0 cm and slit height 0.038 cm at 60 psig average transmembrane pressure drop. Porter shows how the gp model is approximately one order of magnitude too small when compared to the experimental results.

Madsen's (1977) incorrect representation of the lift model at the initial time $t = 0$ is easily explained with Figure 10.19. At $t = 0$, $V_L = 0.436 \times 10^{-5}$ cm/sec^{-1}, whereas $V_{w,\text{initial}} = 0.120 \times 10^{-1}$ cm sec^{-1}, which is far larger by several orders of magnitude. However, the lift velocity V_L increases rapidly while V_w decreases during the experiment, eventually meeting at $V_w = V_L$. It is this "equilibrium" value shown in Figure 10.19 as 8.52×10^{-4} cm/sec for a positive displacement pump and 6.4×10^{-4} cm/sec for a constant power pump that defines the beginning of no deposition. These values when compared to the experiment of Porter (1972) are of the proper order of magnitude. As expected, the "equilibrium" flux for the positive displacement pump is greater than that for the centrifugal pump by about 35%.

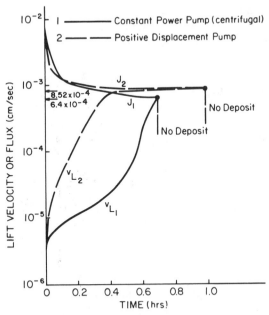

Figure 10.19. Variation of flux and lift velocity obtained from the iterative procedure described in the text for two types of prime movers. Data used here were taken from Figure 31 of Porter (1972) and included $\Delta P = 60$ psi, $\mu = 0.01$ poise, Re $= 1000$, slit height $B = 0.038$-cm, slit width $w = 1.01$ cm, $r_p = 0.1$ μm, $c_f = 10{,}000$ ppm, ϵ 0.25, $J_{\text{initial}} = 1.2 \times 10^{-2}$cm/sec. (Green and Belfort, 1980.)

Although an understanding of the particle dynamics resulting from hydrodynamic forces in the far-field region will be of help in minimizing fouling, many recent studies have suggested that solute–solute and solute–membrane interactions in the near-field region play a vital role in the fouling process (Matthiasson and Sivik, 1980; Dejmek, 1975; Ingham et al., 1980; Dejardin et al., 1980; Howell and Velicangil, 1980; Schilndhelm et al., 1981; Bauser et al., 1982; Charlton et al., 1982; Randerson and Taylor, 1983; Fane et al., 1983a; Fane et al., 1983b; Lange, 1983; Reihanaian et al., 1983; Matthiason et al., 1983; Matthiasson, 1983). Concentration polarization, or the convective transport of retained species toward a permselective membrane, is limited by one of three processes as the solute concentration increases at the membrane–solution interface: gel formation, osmotic and applied pressure equilibration, and solute–membrane adsorption (Belfort and Altena, 1983; Doshi, 1984).

Recent work in the author's laboratory with polystyrene latex and dextrans and with a *Bacillus* fermentation broth confirm the importance of macromolecular adsorption, sometimes an instantaneous process, followed by cell deposition and cake growth. Using 0.35- and 0.55-μm polystyrene latex spheres at different concentrations and axial flow rates in a hollow fiber microfilter, Nicoletti (1985) has confirmed lateral migration effects away from the membrane surface. Herouvis (1984) has shown that at axial flow rates in the laminar regime, cell (*Bacillus polymyxa*) capture at the microfilter and subsequent flux-decline results. Nagata (1985) has shown for the same broth but under turbulent flow conditions a chemical precipitate [probably $Mg(NH_4)PO_4$] resulting from breakdown of urea during sterilization and recombination with nutrients magnesium and phosphate was most likely responsible for the fouling.

From these results and others, adsorption processes must be incorporated into the fouling model. For small organic molecules, adsorption kinetics is fast and probably not rate limiting. For large polymers, however, the kinetics of adsorption is very slow and could influence the flux-time behavior over a long period. Leonard and Vassilieff (1985), Herouvis (1984), and Davis (1985) have suggested that the fouling layers may in some cases move axially along the membrane. More experiments are needed to support this idea.

7. APPLICATIONS

That the worldwide market for membranes was less than $5 million annually in 1960 and has been estimated to be well in excess of $500 million annually in 1981 suggests that membrane technology has arrived. The size of the various segments of the membrane-based industries according to Lonsdale (1982) is shown in Table 10.15.

Applications for membrane filtration are reviewed in Table 10.16, whereas the major specific applications and new and promising applications

TABLE 10.15. Projected Market Size for Various Membrane-Related Industries

Industry	Market size ($ million/year)
Microfiltration	>150
Dialysis, including hemodialysis	200
Electrodialysis	35
Reverse osmosis	100
Ultrafiltration	50
Gas separations, including membrane lungs	10–15
Electrodes	10
Controlled release	>100
All others[a]	> 30

Source: Lonsdale, 1982.

[a] Includes battery separators, chlor-alkali cells, and fuel cells, but excludes packaging film, for example.

TABLE 10.16. Applications for Membrane Filtration

Biotechnology
    Pharmaceuticals (parentals, serum, deionized water)
    Genetic engineering (fractionation, separation)
    Medical diagnostics
    Immunology
    Fermentation (cell harvesting, enzyme clarification)
Agribusiness
    Dairy (effluent treatment)
    Food (starch, sugars)
    Fermentation (wine, dairy)
Industrial
    Recovery of valuable products (paints, dyeing)
Energy
    Fermentation (alcohol, sugars)
Metallurgy
    Precious metal removal and recovery
Leisure
    Sailing
    Camping
Environmental
    Recovery of by-products (valuables)
    Wood pulping (spent sulfite liquor)
    Dairy (whey)
    Municipal (tertiary treatment)
Municipal
    Water supply (desalination of brackish and sea water)

TABLE 10.17. Major Specific Applications for Ultrafiltration[a]

Process	UF Application	Recent Developments	Problems	Competitive Processes
Electrocoat painting	Recovery of anodic colloidal dispersed paint from risings and removal of soluble impurities and corrosion products	Recovery of cathodic dispersed paint from rinsings	Membrane fouling	None
Biological macromolecule production	Purification and concentration of biological macromol	—	High shears may damage macromolecules	Dialysis, ion exchange, lyophilization, selective precipitation
Dairy	Recovery and concentration of protein values from cheese whey	Recovery of lactose for fermentation to alcohol; enzymatic treatment of whey	Fouling and frequent cleaning and sanitizing degrade membranes	—
Electronics and pharmaceuticals	Prepare sterile fluids for medical use and pretreat water to remove colloids prior to demineralization by ion exchange or reverse osmosis	—	—	—
Pyrogen removal or virus concentration	Remove fever-producing (pyrogenic) mucopolysaccharides for medical water, and concentrate virus from surface waters for virus detection (assaying)	Developed automatic hollow-fiber module	Fouling may reduce fluxes	—
Nephrology	Ultrafiltration of blood as artificial kidneys to remove toxic metabolic wastes (hemoultrafiltration)	Coupling UF with dialysis and later infusion of sterile reconstituting fluid	Complex procedure requiring high level of control	Dialysis

Source: After Michaels (1981).

288

TABLE 10.18. New Applications and Promise for Ultrafiltration

Technology	Ultrafiltration Application	Examples	Advantages
Immunochemistry	Preliminary concentration and purification step or fractionation of immunocomplexes from residual haptens. This increases the sensitivity of assay procedures and may also reduce assay time.	Solutes such as antigens, antibodies, peptide hormones in biofluids.	Increases the sensitivity of assay procedures and may also reduce assay time.
Fermentation	Use tangential flow across a membrane to remove the solvent from an aqueous suspension of particulates such as cells, cell debris.	Advantages of using cross-flow filtration of whole fermentation broth are enhanced product recovery, greater product purity, and reduced product isolation costs.	Membrane plasma phoresis to produce a cell filtrate containing a full complement of plasma proteins and an undamaged cell concentrate.
Hydrometallurgical Processing	Selective separation and concentration of trace metals present in tailings streams or leach solutions is desired.	Together with selective chemical complexation techniques, UF can be used to remove specified metals from streams. The metal is then decoupled from the complex.	Allows for the recovery of valuable trace metals, the production of an environmentally acceptable waste stream.
Bioreactors	Hollow fibers could provide the conduit for supplying the nutrients (media, O_2, CO_2) to a fermentation broth either trapped in the understructure of the fibers or free in the ''shell side'' of the reactors. Cells or enzymes could then be immobilized with reactant and products continuously produced and removed from the reactor.	Synthesis of insulin from continuous culture of pancreatic islet cells, continuous production of monoclonal antibodies from culture of hybridoma cells.	Reaction rates could be optimized with the best reactor (nutrient) supply rate. Products could be removed continuously.

Source: After Michaels (1981).

TABLE 10.19. Comparison of Ultrafiltration to Competitive Processes

Process	Advantages	Disadvantages
Adsorption	Simple, easy to scale up	High shears, recovery efficiency low, adsorbent losses high, costly
Ion exchange	Simple, easy to scale up	High shears, regenerants, fouling of resins, costly
Gel permeation chromatography	Simple, good separation	Difficult to scale up; slow, high shear
Electrophoresis and electrofocusing	Excellent separation	Complicated, difficult to scale up; slow, sensitive to changes
Differential migration		
Settling	Simple, easy to scale up	Very slow, relatively inefficient
Centrifugation	Fast	Complicated, difficult to scale up; high shears, expensive
Filtration		
Deep bed	Simple, easy to scale up	High shears, nonselective
Cake	Simple, easy to scale up	Biological cakes very compressive; need inordinate amount of filter aid
Liquid–liquid extraction	Easy to scale up	Expensive, may have solvent losses, very specific process to each case
Ultrafiltration	Excellent separation, relatively easy to scale up; fast, low shear	Fouling of membrane

for ultrafiltration are presented in Tables 10.17 and 10.18, respectively. Membrane technology is not a panacea for all recovery and separation problems. In many cases other processes are either more effecitve, less expensive, or both. However, pressure-driven membrane processes such as reverse osmosis, ultrafiltration, and microfiltration have certain specific and in many cases overwhelming advantages over the choice of other processes. A comparison of ultrafiltration to various competitive processes is given in Table 10.19.

In biochemical applications, often the required product is an intracellular enzyme or protein and is not naturally excreted. *Escherichia coli* is a good example of an intracellular-producing species. The downstream processing for a typical *E. coli* fermenter effluent is shown in Figure 10.20 (Hedman, 1984). After the cells are concentrated by centrifugation (Fig. 10.21a) or tangential flow micro- or ultrafiltration (Fig. 10.21b) they need to be disrupted (Hedman, 1984). This can be done mechanically in a bead mill, chem-

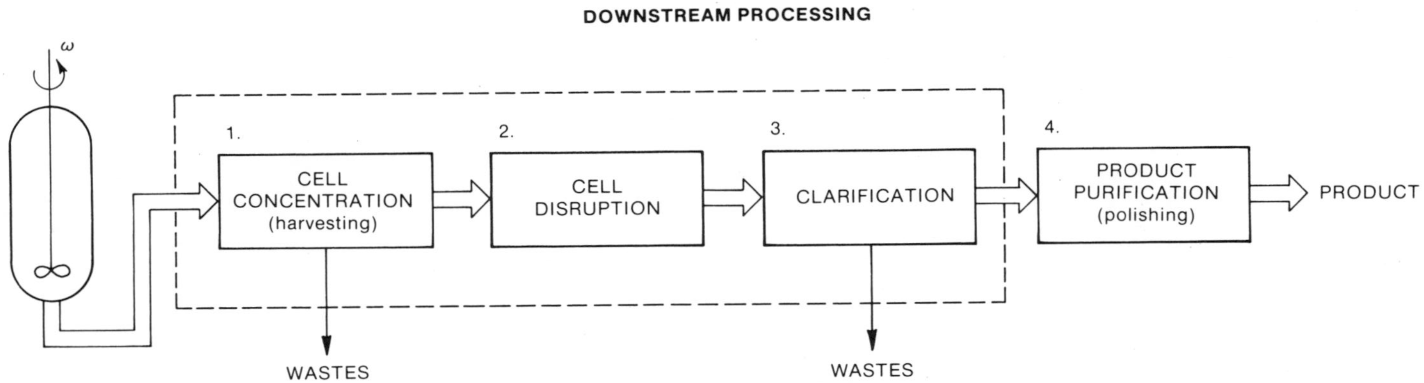

DOWNSTREAM PROCESSING

Figure 10.20. Typical general scheme for downstream processing. (Hedman, 1984.)

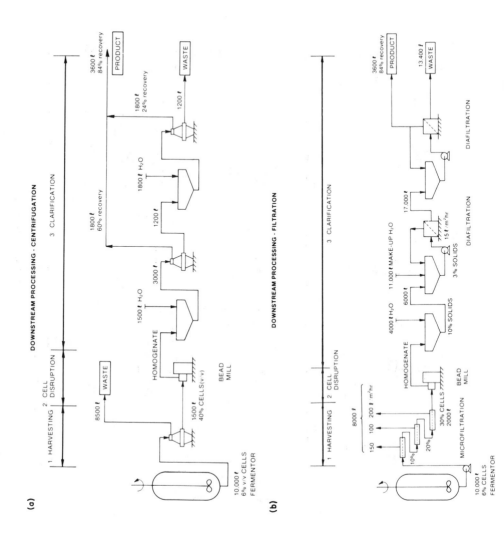

DOWNSTREAM PROCESSING - CENTRIFUGATION

(a)

DOWNSTREAM PROCESSING - FILTRATION

(b)

292

ically, or even biologically. We believe it could also be done inside the cross-flow filter obviating the need for a special cell disruption step. The released product is separated from the cell debris by washing and postcentrifugation or diafiltration. The product is now ready for final polishing usually with chromatographic techniques. In summary, if 15 wt/wt % of E. coli is total protein and 5 wt/wt % of that is the desired product, then with a 84% yield we discover that 0.63 wt/wt % of the original protein has to be recovered. This poses a severe challenge to the separation train and results in the need for exotic processes such as pressure-driven membrane processes and liquid chromatographic techniques.

8. CONCLUSIONS

Besides high-performance membranes, without which the pressure-driven processes discussed here (hyperfiltration, ultrafiltration, and microfiltration) could not be economically competitive, hydrodynamic optimization ranks as one of the most important and continual concerns for maintaining maximal performance. However, in some applications unwanted adsorption processes may ruin performance.

In this chapter we have presented an overview of recent membrane module developments. Transport equations and an analysis of characteristic dimensions have been presented. Both concentration polarization and its component—membrane fouling—have been presented in review and in light of the current knowledge of hydrodynamics in ducts.

We presented the basic problem of fouling in reverse osmosis ultrafiltration and microfiltration and have critically evaluated the explanations proposed for this phenomenon. A more consistent general model for fouling has been presented that demonstrates that previous models, once thought to be mutually exclusive, are actually mutually supportive components of the general theory. We have also suggested that the mechanisms of shear, lift, convective drag, adsorption, and cake morphology are at work throughout a process and can explain steady, transient, and long-term effects.

ACKNOWLEDGMENTS

The author would like to thank his students, Gordon Altshuler, George Green, Jeff Schonberg, Roger Weigand, Cheng-sheng Lee, Jim Nicoletti, Kay Herouvis, Naohiko Nagata, and Jane Otis; visiting scholar Dr. Frank Altena; staff biologist Dr. David Dziewulski; and technician Thomas

Figure 10.21. Detail flowsheet of downstream processing with centrifugation (a) replaced by (b) membrane filtration. (Hedman, 1984.)

Neubauer for their significant contributions to the ideas developed in our laboratory and presented here.

REFERENCES

Aerojet General Corp., U.S. Off. Saline Water, Res. Develop Progr. Rep. 86, 1964.

Aerojet General Corp., U.S. Off. Saline Water, Res. Develop Progr. Rep. 213, 1966.

Ajax International Corp., Sales New Flash 73-10-1, 1973, p. 1.

Altena, F. W., and Belfort, G. *Chem. Eng. Sci.* **39**(2), 343–355, 1984.

Altena, F. W., Belfort, G., Otis, J., Fiessinger, F., Rovel, J., and Nicoletti, J. *Desalination* **47**, 221–232, 1983.

Bauser, H., Chmiel, H., Stroh, N., and Waltza, E. *J. Memb. Sci.* **11**, 321, 1982.

Belfort, G. Pressure-driven membrane processes and wastewater renovation, in *Water Renovation and Reuse,* New York, Academic Press, 1977, Chap. 6, pp. 129–189.

Belfort, G. Desaling experience by hyperfiltration (reverse osmosis) in the U.S.A., in Belfort, G. (ed.), *Synthetic Membrane Processes, Fundamentals and Water Applications,* New York, Academic Press, 1984, Chap. 7, pp. 221–280.

Belfort, G., and Altena, F. W. *Desalination* **47**, 105–127, 1983.

Belfort, G., and Marx, B. (1979). Artificial Particulate Fouling of Hyperfiltration Membranes— II Analysis and Protection from Fouling, Desalination, **28**, 13–30, 1979.

Belfort, G., Chin, P., and Dziewulski, D. M. New Gel-Polarization Model Incorporating Lateral Migration for Membrane Fouling, Proceedings of World Filtration Conf., Philadelphia, The Filtration Society, London, U.K., Sept. 1982.

Belfort, G., Littman, F., and Bishop, H. K. *Water Res.* **7**, 1547, 1973.

Bennion, D. N., and Rhee, B. W. *Ind. Eng. Chem. Fund.* **8**, 36, 1969.

Blatt, W. F., Drauid, A., Michaels, A. S., and Nelson, L. In Flinn, J. E. (ed.), *Membrane Science and Tech.,* New York, Plenum, 1970, p. 47.

Bokhorst, R., Altena, F. W., and Smolders, C. A. Proceedings of IDEA Conference, Bahrain, 1981.

Bray, D. T. U.S. Patent 3,417,870, 1968.

Breton, E. J., Jr., and Reid, C. E. *AIChE Chem. Engr. Prog. Symp., Ser.* **55**, No. 24, 171, 1959.

Bruin, S., Kikkert, A., Weldrig, J. A. G., and Hiddink, J. *Desalination* **35**, 223–242, 1980.

Cadotte, J. E., and Petersen, K. J. In Turbak, A. F. (ed.), *Synthetic Membranes, Desalination.* Washington, D.C. ACS Symposium Series 153, ACS, 1980, vol. 1, Chap. 21, pp. 305–326.

Charlton, B., Schindhelm, K., and Farrell, P. C. *Trans. Am. Soc. Artif. Internal Organs* **28**, 400–403, 1982.

Davis, R. H. Proceedings of Physicochemical Hydrodynamics V, Tel Aviv, Israel, Dec. 17–21, 1985.

Dejardin, P., Toledo, C., Pefferkorn, E., and Varoqui, R. Flow rates of solutions through ultrafiltration membranes monitored by the structure of adsorbed flexible polymers, in Copper, A. R. (ed.), *Ultrafiltration Membranes and Applications,* New York, Plenum Press, 1980.

Dejmek, P. Concentration polarization in ultrafiltration of macromolecules, Ph.D. Thesis, Lund Institute of Technology, Lund, Sweden, 1975.

Doshi, M. R. Limiting Flux in UF of Macromolecular Solutions, presented at the Annual 1984 ACS Meeting, Philadelphia, Aug, 26–30, 1984.

Eckstein, E. C., Bailey, D. G., and Shapiro, A. H. *J. Fluid Mech.* **79**, 191, 1972.

Fane, A. G., Fell, C. J. D., Suki, A. *J. Memb. Sci.* **16**, 195, 1983a.

Fane, A. G., Fell, C. J. C., Waters, A. G. *J. Mem. Sci.,* **16**, 211, 1983b.

Ferry, J. D. *Chem. Rev.* **18**, 373, 1936.

Fick, A. *Uber Diffusion, Pogg. Ann.* **94**, 59, 1865.

Frommer, M. A., and Lancet, D. (1972). In Lonsdale, H. K., and Podall, H. E. (eds.), *Reverse Osmosis Membrane Research,* New York, Plenum Press.

Graham, T. *Phil. May* **32**, 402, 1866.

Green, G., and Belfort, G. *Desalination* **35**, 129–147, 1980.

Grover, J. R., Gaylor, R., and Delve, M. H. Proc. Int. Symp. Fresh Water Sea, 4th, 1973, vol. 4, pp. 159–169.

Gutman, K. G. *Chem. Engr. (U.K.)* **322**, 510, 1977.

Hasse, R. *Thermodynamics of Irreversible Processes.* Reading, MA. Addison-Wesley, 1969.

Hansmann, G., and Pietsch, J. *Naturwissen Schaften* **36**, 250, 1949.

Hedman, P. *Am. Biotech. Lab.* **2**(3) 29–39, 1984.

Helmcke, J.-G. *Kolloid Z* **135**, 101, 1954.

Hendricks, T. J., and Williams, F. A. Boundary Layer Flow Problems in Desalation by Reverse Osmosis, OSW Final Report, Grant No. 14-01-0001-951, June 1970.

Herouvis, K. M.S. Thesis, RPI, Troy, NY 12180, 1984.

Ho, B. P., and Leal, L. G. *J. Fluid Mech.* **22**, 385, 1965.

Howell, J., Velicangil, O., Protein ultrafiltration: Theory of membrane fouling and its treatment with immobilized proteases, in Cooper, A. R. (ed.), *Ultrafiltration Membranes and Applications,* New York, Plenum Press, 1980, p. 217.

Ingham, K. C., Busby, T. F., Sahlestrom, Y., Castino, F., Separation of macromolecules by ultrafiltration: Influence of protein adsorption, protein-protein interactions, and concentration polarization, in Cooper, A. R. (ed.), *Ultrafiltration Membranes and Applications,* New York, Plenum Press, 1980, p. 141.

Ishi, K., and Hasimoto, H. *J. Phys. Soc. Jpn.* **48**(6), 2144–2155, 1980.

Jonsson, G., and Boessen, C. E. In Belfort, G. (ed.), *Synthetic Membrane Processes, Fundamentals and Water Applications,* New York, Academic Press, 1984, chap. 4, pp. 102–130.

Juda, W., and McRae, W. A. *J. Am. Chem. Soc.* **72**, 1044, 1950.

Kamiyama, Y., Yoshioka, N., Matsui, K., and Nakagome, K. *Desalination* **51**, 79–92, 1984.

Kleinstreuer, C., and Belfort, G. In Belfort, G. (ed.), *Synthetic Membrane Processes, Fundamentals and Water Applications,* New York, Academic Press, 1984, Chap. 5, pp. 131–190.

Kolf, W. J., and Berk, H. T. *Acta Med. Scand.* **117**, 121, 1944.

Kozinski, A. A., Schmidt, F. P., and Lightfoot, E. N. *Ind. Eng. Chem. Fund.* **9**(3), 502–505, 1970.

Kurihara, M., Harumiya, N., Kanamaru, N., Tonomura, T., and Nakasatomi, M. Desalination **38**, 449, 1981.

Lange, K. E. Ph.D. Thesis in Chemical Engineering, Stanford University, Stanford, CA, 1983.

Lee, H. C., and Lightfoot, E. N. *AIChE Journal* **20**(2), 335–339, 1974.

Leonard, E. F., and Vassilieff, C. S. *Chem. Engr. Comm.,* 30 (3–5) 209–217, 1984.

Loeb, S., and Sourirajan, S. *Adv. Chem. Ser.* **38**, 117, 1982 and U.S. Pat. 3,133,132, May 12, 1964.

See Lonsdale, H. Review in *J. Membrane Sci.* **10**, 81, 1982.

Lonsdale, H. K. *J. Mem. Sci.* **10**, 81–181, 1982.

Madsen, R. F., *Hyperfiltration and Ultrafiltration in Plate-and-Frame Systems,* Amsterdam, Elsevier Scientific Publishing, 1977.

Madsen, R. E., *Hyperfiltration and Ultrafiltration in Plate and Frame Systems,* Amsterdam, Elsevier, 1977.

Madsen, R. F., Olsen, O. J., Nielsen, I. K., and Nielsen, W. K. Use of hyperfiltration and ultrafiltration with chemical and biochemical industries, in Linder, G. and Nyberg, K. (eds.), *Environmental Engineering, A Chemical Engineering Discipline,* Dordrecht, Netherlands, Reidel, 1973, pp. 320–330.

Mahlab, D., Ben Yosef, N., and Belfort, G. (1980) *Chem. Engr. Commun.* **6,** 225–243, 1980.

Mahlab, D., Ben Yosef, N., and Belfort, G. ACS Symp. Series No. 153, in Tarbak, A. F. (ed.), Synth. Membranes: Desalination CS, Washington, 1981, vol. 1, Chap. 10, pp. 148–158.

Mahon, H. I. *National Research Council Publication* **942,** 345, 1963.

Mahon, H. I., McLain, E. A., Skiens, W. E., Green, B. J., and Davis, T. E. *AJChE Chem. Engr. Prog. Symp. Ser.* 91, **65,** 48, 1969.

Matthiasson, E., *J. Memb. Sci.* **16,** 23, 1983.

Matthiasson, E., and Sivik, B. *Desalination* **35,** 59–103, 1980.

Matthiasson, E., Hallstrom, B., and Sivik, B., Engineering and Food, **1,** Proc. from the Third International Congress on Engineering and Food, Dublin, Ireland, 1983.

Mertin, U. In Merten, U. (ed.), *Desalination by Reverse Osmosis,* Cambridge, MA, The MIT Press, 1966, Chap. 2.

Merten, U. *Desalination by Reverse Osmosis,* Cambridge, MA, MIT Press, 1966.

Meyer, K. H., and Sievers, J. F. *Helv. Chim. Acta.* **19,** 649, 1936.

Michaels, A. S. *Chem. Eng. Prog.,* **64,** 31, 1968.

Michaels, A. S. Chemtech 36–43, January, 1981.

Nagata, N. Unpublished, RPI, Troy, NY 12180, 1985.

Nakao, S., and Kirmura, S. In Turbak, A. F. (ed.), *Synthetic Membranes,* ACS Symposium Series 154, Warbuytin, ACS, 1981, vol. II, Chap. 9.

Nicoletti, J. Unpublished, RPI, Troy, NY 12180, 1985.

Nielson, W. K. The Use of Ultrafiltration and Reverse Osmosis in the Food Industry and for Wastewaters from the Food Industry, Paper No. WKN/1h. Obtainable from DDS, Nakshov, Denmark, 1972.

Onsager, L. *Phys. Rev.* **37,** 405, 1931.

Porter, M. C. *I & EC Prod. Res. Dev.* **11** 234, 1972.

Prigogine, I., *Introduction to Thermodynamics of Irreversible Processes,* Springfield, Illinois, Thomas, 1955.

Probstein, R. F., Leung, W. F., and Allianc, Y. *J. Phys. Chem.* **83**(9) 1230, 1979.

Pusch, W., and Walch, A. *Angen. Chem. Int. Ed. Engl.* **21,** 660, 1982.

Pusch, W., and Woermann, D. *Ber. Bunsenges, Physik Chem.* **74,** 444, 1970.

Randerson, D. H., and Taylor, J. A. Protein adsorption and flux decay in membrane plasma separators, in Nosé, Y. et al. (eds.), Proc. 2nd Intl. Symp. on Plasmapheresis, Cleveland, 1983.

Reed, R. H., and Belfort, G. *Water Sci. Tech.* **14,** 499–522, 1982.

Reihanaian, H., Robertson, C. R., and Michaels, A. S. *J. Mem. Sci.* **16,** 237, 1983.

Riley, R. L., Lonsdale, H. K., and Lyons, C. R. *J. Appl. Polym. Sci.* **15,** 1267, 1971.

Rozelle, L. T., Cadotte, J. E., King, W. L., Senechal, A. J., and Nelson, B. R. OWS Research Progr. Report No. 659, 1973.

Sachs, B., Shelef, G., and Ronen, M. Renovation of Municipal Effluents by Sewage Ultrafiltration, Dept. of Membrane Processes, Israel Desalination Engineering, Tel Aviv, Israel, 1975.

Schilndhelm, K., Roberts, C. G., and Farrell, P. C. *Trans. Am. Soc. Artif. Internal Organs* **27**, 554–558, 1981.

Segre, G., and Silberberg, A. *J. Fluid Mech.* **14**, 115, 1962.

Segre, G., and Silberberg, A. *Nature,* **189**, 209, 1961.

Sherwood, T. K., Brain, P. C. T., Fisher, R. E., and Dresner, L. *Ind. Engr. Chem. Fundamentals* **4**, 113, 1965.

Spiegler, K. S., and Kedem, O. *Desalination* **1**, 311, 1966.

Strathmann, H., in Belfort, G. (organizer), International Symposium on Membranes and Wastewater Treatment, Hebrew University, Jerusalem, Israel, 1973.

Strathmann, H. in Belfort G. (ed.), *Synthetic Membrane Processes, Fundamentals and Water Applications,* New York, Academic Press, 1984, chap. 9, pp. 343–375.

Strathmann, H., Kockk, Amar, P., and Baker, R. W. *Desalination* **16**, 179, 1975.

Staverman, A. J. *Rec. Trav. Chim., Pays-Bas Bldg.* **70**, 344, 1951.

Teorell, T. *Proc. Soc. Expt. Biol.* **33**, 282, 1935.

Thomas, D. G., and Gallaher, R. B., Hydrodynamic Flux Control for Waste Water Application of Hyperfiltration Systems, Oak Ridge National Laboratory, EPA report No. EPA-R2-73-228, May 1973.

Thomas, D. G., Gallaher, R. B., and Johnson, J. S., Jr. Hydrodynamic flux control for wastewater application of hyperfiltration systems. Environ. Protet. Technol. SEr. EPA-R2-73-228, 1973.

Trettin, D. R., and Doshi, M. R. *Ind. Eng. Chem. Fund.* **19**, 184–194, 1980.

Trettin, D. R., and Doshi, M. R. In Turbak, A. F. (ed.), *Synthetic Membranes,* ACS Symp. Series No. 154, Washington, D.C., 1981, chap. 22.

Westmoreland, J. U.S. Patent 3,367,504, 1968.

Wilson, J. R. (ed.) *Dimineralization by Electrodialysis,* Buttersworth, London, 1960.

Zsigmondy, R., and Bachman, W. *Z. Anorg. Chem.* **103**, 119, 1918.

Zydney, A. L., and Colton, C. K. Continuous Flow Membrane Amapheresis: Theoretical Models for Flux and Hemosysis Prediction, ASAIO annual meeting, Chicago, 1982.

INDEX

Absorption, using polysaccharides, 220
Acetic acid, 9
Acetoin, 9
Acid hydrolysis, 85, 86
Adsorption:
  productivity, 231–234
  solute onto membrane, 286
Agarose gel, 114, 117
Agricultural chemicals, 6, 8
Alcohol dehydrogenase, 151
Alcohols, 5
Algae, removal, 208, 210
Allosteric regulation, 125
Amensalism, 25
Amino acid, 4
Androstenedione, 226
Anneal by hydrogen bonding, 106
Antibiotics, 6, 7, 167–169, 171, 172, 174, 175, 178, 180, 182, 183
Aqueous partitioning, 213
Ascorbic acid, 6
Aspergillus, 175, 176
Autonomous replicating sequence (ARS), 112
Autoradiography, 117
Auxotrophic mutant, 116

Bacillus amyloliquefaciens, 140
Biomass refining, cost estimates, 100
Bioreactor:
  classification systems, 37
  design modeling equations, 44
  simulation models, 41
Blunt-end ligation, 104, 108
Butanediol, 2, 3, 11
2,3-Butanediol, 94
Butanol, 5

Capacity factor, 225
Carbapenem, 172
Cell debris, 189
Cellulose:
  availability, 80, 81
  hydrolysis, 89
  models, 86
  structure, 82, 84
Centrifugation, 198
Cephalosporin, 169, 170, 178, 182
Cephalosporium, 169, 182
Cerulenin, 172, 173
Chemical composition of wood, 93
Chemical intermediates, 2, 8
Chemical modification, 143
Chimeras, 103, 106
Chloramphenicol, 178
Chromatography, 210–213
  adsorption, 211
  diffusity, 212
  gel filtration, 212
  ion exchange, 212
  liquid-liquid, 212
Chymotrypsin, 132
  enzyme mimic, 160
Citric acid, 4
Cloned DNA, 104
Cloned yeast centromere, 113
Cloning, 103, 183
Coagulation:
  electrostatics, 194–197
  kinetics, 197–198
  stability, 194
Colicinogenic plasmic colE1, 110
Column length, 229
Competition, 24, 25

Complementary DNA, 104, 106
Complementation, 117
Computer control applications, 42
Computer simulations, 52
Concentration polarization, 265–286
    chemical and physical membrane methods,
        269
    designs to reduce, 267, 268
    limitations, 265, 266
    measurement, 265, 266
Concentration processes:
    chromatography, 210–213
    coagulation and flocculation, 194–198
    differential migration, 198–200
    electrically driven, 213
    filtration, 201–210
    precipitation, 191–193
Continuous fermentation, 11
Copy number, 112
Corn grits, 221
Cosmids, 112
Cost estimates, 100
Cost of support, 223
Cyanogen bromide, 122
Cyclodextrins, enzyme mimics, 158

Daunorubicin, 172
Deep shaft fermenter, 33
Defective recombinant SV40, 113
Degree of polymerization, 91
Deletion mutations, 124
Deregulation, 19
Derjaguin–Landau–Verwey–Overbeek
        (DLVO) theory, 195, 201
Desorption, 221
Diffusion coefficient, 228, 229
Directed biosynthesis, 19
Direct fermentation of cellulosic biomass, 98
Distillation, 4, 11
    of ethanol, 220
Distributor, 223
Diversity, 24
DNA, 180, 182, 183
    binding-protein, 156
    ligase, 105
    manipulation, 7
    phase M13, 125
    polymerase I, 106
    RNA hybrids, 115
Downstream processing, 189, 190
Drug-resistance markers, 116

E. coli, 142
Effectiveness factor, 30

EK2 containment hosts, 108
Elective cultures, 23
Electrodialysis, 213, 241
Electrophoresis, 114
Eluant cost, 235
Embden–Meyerhof pathway, 9
Enzymatic hydrolysis, 85, 95
Enzyme(s), 7
    activity:
        in organic solvents, 150
        on unusual substrates, 151
    advances in analytical techniques, 144
    advantages, 130
    alcohol dehydrogenase, 151
    alteration by genetic engineering, 136
    alteration of substrate specificity by genetic
        engineering, 140
    catalysis, 132
    chemical modification, 144
    chymotrypsin, 132, 160
    computer graphics, 144
    denaturization, 29, 192
    extracellular, 189
    forces stabilizing conformation, 133
    halophilic, 154
    intracellular, 189
    lactate dehydrogenase, 148
    lipase, 146, 150
    lowering K_m by genetic engineering,
        138
    melittin, 155
    nonpeptide mimics, 157
    novel catalytic activities, 149
    oxidoreductases, 151
    oxygenases, 152
    problems, 153
    properties amenable to modification, 143
    reaction types, 149
    recovery, 188
    ribonuclease S, 141
    structure and function, 132
    subtilisin BPN', 140
    synthetic peptide mimics, 154
    thermolysin, 147
    thermophilic, 154
    tyrosyl tRNA synthetease, 137
Equations of motion, 269
Erythromycin, 173
Escherichia, 180
Ethanol, 2, 94
Eukaryotic cytoplasmic mRNAs, 114
Expression:
    of cloned eukaryotic genes, 108
    of eukaryotic genes in prokaryotes, 109

Ferrocene, enzyme mimic, 159
Filtration:
  attachment, 201
  cake, 208
  deep bed, 201–208
  high gradient magnetic filtration, 208
  membrane, 209
  transport, 201–208
Flocculants, 192–193
Flocculation:
  orthokinetic, 197–198
  perikinetic, 197
Fluid flow on porous ducts, 269–271
Fluidized reactors, 36
Foods, 2
Foulants, 267, 286
Fouling, 266–286
Fructose, 4, 7
Fuels, 2
Furfural, 92
Fusion peptide, 120

β-Galactosidase protein, 120
Gas-liquid bioreactor, 52
Gel polarization:
  lateral migration model, 279–281
  model, 272–279
Gene:
  cloning, 174, 175
  fusion, 120
  library, 114
  synthesis, 142
Genetic code, 115
Genetic engineering, 6, 9, 103, 168
  in vitro synthesis of genes, 142
  site-directed mutagenesis, 136
Genetics, 167, 168, 172
Glucose, 3, 7, 9
Glucose-isomerase, 94
Glycerol, 5, 9
G protein, 122
Growth:
  limitation, 22
  spatial effects, 14

Hemicellulose, 81
Hemicellulose hydrolysis, 87, 92
Heper, 113
Highly repetitive DNAs, 114
High plate count, 225
Hollow fiber biochemical reactor:
  case study, 68
  substrate uptake, 45
Hormones, synthetic mimics, 156

Hybridization, 116
Hyperfiltration (reverse osmosis), 11, 240, 241,
  251, 252, 255, 256, 258–260, 263, 265

Ideal conditions, 229
Immobilization techniques, 123
Immobilized biocatalyst reactors, 16, 67
Immobilized enzymes, 26, 27
Initiation codon, 115
Insertions, 124
Insulin genes, 122
Interactions, membrane-foulant, 267, 268
Interferon, 7
Interferon (INT) genes, 123
Intermediates, 5
Intron-exon, 126
In vitro mutagenesis, 124
In vitro packaging system, 110
Itaconic acid, 5

Kanamycin, 178–180, 182

Lac operon promoter, 120
β-Lactam, 169, 171
β-Lactamase, 171
Lactate dehydrogenase, 148
  chemical modification, 148
Lactic acid, 5, 9
Lambda pL regulatory region, 120
Lateral migration (or particles), 279–286
Lignin, 84
Lincomycin, 173, 174
Linear chromatography, 228
Lipase:
  activity in organic solvent, 150
  polyethylene adducts, 146
Liquid chromatography (LC), 219

Macrolide, 173
Maridomycin, 173
Mass transfer correlations, 275
  cross-flow (continuous), 272–286
  laminar flow, 275, 276
  turbulent flow, 275, 277
  unstirred batch, 271–272
Material balance, 229
Melittin:
  synthetic mimic, 155
Membrane permeators:
  comparison, 259–260, 261
  fiber, 257–258
  helical and rigid tube, 254–256
  modules, 248–260
  plate and frame, 257–258

Membrane permeators (*Continued*)
  spiral wrap, 254–255, 257–258
Membrane processes:
  applications, 286–293
  classification, 242, 244
  downstream processing, 290–293
  history, 239–240
Membranes, 242, 248
  annealing, 244, 248
  assymmetric, 240
  classification, 242, 243
  composite, 244, 245
  criteria for commercialization, 242
  finely porous, 244
  formation, 247
  homogeneous, 244
  market size, 287
  mechanism of formation, 244, 246
  methods for characterizing, 244, 245
  new materials, 243, 244
  solution-diffusion, 244
  transport equations, 244, 260–265
  type of material, 244–250, 251–253
Membrane separation, 11
Mesophilic, 125
Metabolic control, 17
Metabolic paths, 9
Methane, 2
Methanol, 4
Methyl ethyl ketone, 94
Microfiltration, 240, 241, 265, 290–293
Mixed cultures, 21
Mixed reactor trains, 62
Modest plate count, 227
Module(s), 248–260
  requirements, 248, 250
  see also Membrane permeators
mRNA hybridization, 114
Mutagenesis:
  cassette, 140
  classical, 137
  site-directed, 137
Mutants, 168–173
Mutasynthesis, 19, 20
Mutation, 7, 124
Myoglobin, 145

Nitrocellulose paper, 114
Nocardia, 170, 171
Norcardicin, 170, 171
Nonstringent conditions, 125
Nucleation and growth, 246

Operational life, 229, 233

Optimization, 19
Organic intermediates, 5
Oxygenases, activity on unnatural substrates,
  152

Packaging lambda DNA, 110
Packed bed reactors, 36
Pancreatic deoxyribonuclease (DNase), 124
Papain, 145
Parasexual, 175–177
Particle size, 222
pBR322, 110
Penicillin, 169, 170, 176, 178
Penicillium, 169, 175, 182
Permeation, linear flux laws, 243
Pervaporation, 11
Phage lambda vectors, 110
Pharmaceuticals, 2, 6
Phase diagram, 246
Phase-separation, 244–246
Phenotype, 117
Plasmaphoresis, cross-flow, 277, 279, 280
Plasmids, 109, 175, 178, 180–183
  2-μm, 112
Plate count, 224
Plate height, 225, 230
Poly A tails, 114
Polyethylene glycol adduct, 146
Polysaccharides, 221
Porous plate, 223
Precipitation, salting out, 192
Predation, 25
Preparative scale columns, 223
Pretreatment, 89
Primer, 117
Process eukaryotic messenger ribonucleic acids
  (RNAs), 108
Promoter, 118
Proteins:
  DNA-binding, 156
  myoglobin, 145
  papain, 145
  polyethylene glycol adducts, 146
  synthetic enzyme mimics, 154
  see also Enzyme(s)
Protoplast fusion, 172, 177, 178, 183
Protoplasts, 172, 175, 177, 183
Prototrophic colonies, 117
Pseudomonas, 168
Purification, 220
Pyrrolnitrin, 168, 169
Pyruvic acid, 9

Rabies virus, 122

R-DNA, 180, 182, 183
Reading frame, 120
Rear-field region, 286
Rearrangement, 114
Recognition sequence, 104
Recombinant deoxyribonucleic acid (DNA),
    103, 168, 180, 183
Recombination, 114, 175
Recovery processes, 187, 188, 192
    carbon adsorption, 191
    ion exchange, 191
    precipitation, 189
    solvent extraction, 189, 191
Recycle fermentation, 188
Recycle of live microbes, 9
Regenerant cost, 235
Regression analysis, 41
Relative motion of cells and medium, 16
Relaxed replication control, 109
Replication origin, 112
Resistance model, 282–286
Resolution, 227
Restriction endonuclease, 103
Retention volume, 223
Reverse transcriptase, 113
Ribonuclease A, 154
Ribonuclease S, 141
Ribosomes, 118, 120
Roasting-leaching process, 87–89

Saccharomyces, 112, 170
Sample injection, 228
Sample volumes, 232
Sarex process, 220
Scale up, 236
Screening techniques, 116
Secondary metabolite, 17, 18
Sedimentation, 198–200
    lamellar, 200, 204
    settling curves, 202–203
    type of settling, 199
Selection, 26, 104, 110, 168, 172, 183
Selling price, biologicals, 187
Sense transcript, 118
Separation factor, 227
Serine proteases, 132
Shine–Delgrano sequence, 120
Shotgun cloning, 113
Shuttle vectors, 113
Signal sequence, 120
Single-cell protein, 4, 93
Site-directed mutation, 124, 136
Sizing procedure, 117
Solid-state fermentation, 36

Solute-solute interactions, 286
Solvent cost, 235
Solvent refining of biomass, 97
S1 nuclease, 115
Southern blotting technique, 114
Specialty chemicals, 6
Spheroplasts, 112, 116
Splicing, 104
Staphylococcus, 171
Steam explosion, 97
Sticky end splicing, 104
Stirred tank power requirement, 45
Stirred tank reactors, 33, 61
    with recycle, 66
Stoichiometric limitation, 22
Strain development, 18
Strain improvement, 168, 170, 171
Streptomyces, 169, 173, 175, 178, 179, 183
Strong promoters, 120
Substitution mutations, 125
Substrate uptake in biochemical reactor, 45
Subtilisin BPN', site-directed mutagenesis,
    140
Succinic acid, 5, 9
Sucrose gradient sedimentation, 117
Surface glycoprotein, 122
SV40, 113
Synthetic DNA oligonucleotide probes, 117
Synthetic gene, 106, 113
Synthetic oligonucleotides, 108
Synthetic peptide enzymes, 154

Tandemly repeated DNAs, 114
T_4 DNA ligase, 108
Terminal deoxynucleotidyl transferase, 106
Termination codon, 115
Thermodynamics of irreversible processes, 242,
    260–265
Thermolysin, polyethylene adducts, 147
Thermophilic microorganisms, 125
Thymidine kinase deficient (tk−) host, 113
Tower and loop bioreactors, 61
Transfection with viral DNAs, 116
Transformation techniques, 112
Transformed host, 104
Translation, 118
Transport equations, 263–265
Trp-lac regulatory sequences, 120
Tylosin, 174
Tyrosyl tRNA synthetease, site-directed
    mutagenesis, 137

Ultrafiltration, 240, 241, 253, 260, 261, 263,
    288–290, 293

Value balance, 47
Vectors, 104, 109
Viruses, 109
Viscosity, 222, 230
Void fraction, 26, 223

Western block procedure, 124
Whey, 4
Wilke–Chang equation, 228

Wood chips, 3

Xylose, 3, 94
Xylose isomerase, 94

Yeast, 4
Yeast-cloning vectors, 112

Zeta potential, 201